Prepublication endorsements of
Facing up to Climate Reality: Honesty, Disaster and Hope

"It is easy to interpret these fascinating essays as leaning towards the gloomier terrain occupied by the 'Dark Mountain'. But whilst collectively the authors dispense with techno-utopian approaches to mitigation, their development of transformative adaptation provides a more informed response to the challenges we face. Such realism is timely in a world still fixated on fossil fuels and on an emission trajectory toward a four (or more) degrees future. From the ashes of the old hope, *Facing Up to Climate Reality* opens up space for a new and potentially more fruitful dialogue. Whilst the authors clearly recognise our profound crisis, they suggest that we may yet navigate our way through it, and in so doing could deliver improvements in some key aspects of our quality of life. I can only hope that their measured optimism helps catalyse widespread and meaningful action."

— **Professor Kevin Anderson**, Tyndall Centre
for Climate Change Research

"Readers of this fascinating book are brought closer to the truth that world temperatures will rise this century to levels only experienced on the planet many millions of years ago, long before *homo sapiens*; and that catastrophe is therefore almost-inevitably impending."

— **Mayer Hillman**, father of carbon rationing and author
of *How We Can Save the Planet*

"This important new collection brings the trademark radicalism of Green House to the climate crisis. The authors set out an array of bold and hopeful ideas, consider how facing up to climate disasters can kindle new green shoots of community, and explore the psychology of climate communication. The book both pursues climate honesty rigorously and offers hope for the future."

— **Caroline Lucas MP**, author of *Honourable Friends?*
Parliament and the Fight for Change

FACING UP TO CLIMATE REALITY

Green House is a think tank founded in 2011. It aims to lead the development of green thinking in the UK.

Politics, they say, is the art of the possible. But the possible is not fixed. What we believe is possible depends on our knowledge and beliefs about the world. Ideas can change the world, and Green House is about challenging the ideas that have created the world we live in now, and offering positive alternatives.

The problems we face are systemic, and so the changes we need to make are complex and interconnected. Many of the critical analyses and policy prescriptions that will be part of the new paradigm are already out there. Our aim is to communicate them more clearly, and more widely.

We are independent of political parties, campaigns or commercial vested interests, but will be happy to cooperate with any individual or organisation sharing our beliefs and our sense of urgency.

FACING UP TO CLIMATE REALITY

Honesty, Disaster and Hope

Edited by John Foster

GREEN HOUSE   GREENHOUSE
              THINK TANK

LONDON PUBLISHING PARTNERSHIP

Copyright © 2019 by Green House

Published by Green House Publishing
www.greenhousethinktank.org

Published in association with London Publishing Partnership
www.londonpublishingpartnership.co.uk

ISBN 978-1-907994-92-0 (pbk.)
ISBN 978-1-907994-93-7 (iPDF)
ISBN 978-1-907994-94-4 (epub)

A catalogue record for this book is available from the British Library

This book has been composed in Adobe Garamond Pro and
printed on FSC certified paper using vegetable inks

Copyedited and typeset by T&T Productions Ltd, London
www.tandtproductions.com

CONTENTS

Editorial Foreword
John Foster   ix

Notes on Contributors   xiii

Introduction: Looking for Hope between Disaster and Catastrophe
Brian Heatley, Rupert Read and John Foster   1

Part I: Politics   **13**

Chapter 1 – Could Capitalism Survive the Transition to a Post-Growth Economy?
Richard McNeill Douglas   15

Chapter 2 – Facing Up to Climate Reality: International Relations as (Un)usual
Peter Newell   35

Chapter 3 – Making the Best of Climate Disasters: On the Need for a Localised and Localising Response
Rupert Read and Kristen Steele   53

Part II: Systems   **69**

Chapter 4 – Linking Cities and the Climate: Is Urbanisation Inevitable?
Jonathan Essex   71

Chapter 5 – Dealing with Extreme Weather
Anne Chapman   93

CONTENTS

Chapter 6 – Geoengineering as a Response to the Climate Crisis: Right Road or Disastrous Diversion?
Helena Paul and Rupert Read   109

Part III: Framings   **133**

Chapter 7 – What the Crisis of the Late Middle Ages in Europe Can Tell Us about Global Climate Change
Brian Heatley   135

Chapter 8 – Facing Up to Ecological Crisis: A Psychosocial Perspective from Climate Psychology
Nadine Andrews and Paul Hoggett   157

Chapter 9 – Where Can We Find Hope?
John Foster   175

Coda—Where Next?
John Foster on behalf of the Green House Collective   193

Bibliography   199

Sponsors   223

EDITORIAL FOREWORD

John Foster

This book is a sequel to Green House's earlier collection *The Post-Growth Project* (Blewitt and Cunningham, 2014), which sought to get countries like Britain to recognise the vital need to end 'economic growth' in the interests both of human well-being and of ecological stability. In this new book, we begin a closely related but even more challenging task. That is to confront the brutal reality of the long-term climate damage which such growth has already made inevitable. Honesty about this situation is something which green politics (never mind conventional politics) has hitherto found desperately hard, but at the stage which we have now reached it is the necessary prequel to any *effective* climate action.

Authoritative science, from the Intergovernmental Panel on Climate Change (IPCC, 2018) and others, has already documented our looming climate plight, and powerful advocacy such as that of Naomi Klein (2014) and George Monbiot (2017) has urged last-minute action to avoid it. We accept the science, but do not seek to duplicate that advocacy. Instead, we begin the more difficult work of facing up to climate reality—the reality of what our present situation means in practice. What is actually involved in accepting the inescapability, now, of real climate-driven disasters world-wide and the end of the dream of uninterrupted human 'progress'? Seeking answers to these questions, we probe the barriers to the transformative adaptation which we now so urgently need. We recognise that the world is already committed to escalating climate chaos, but yet, at the same time, we refuse to give up hope of avoiding climate catastrophe.

We have grouped nine separately-authored chapters which start to explore the implications of this position under three broad headings—*Politics*, *Systems* and *Framings*—corresponding respectively to the ideological and governance challenges, the huge agenda of practical difficulties and the kinds of intellectual and imaginative regrouping which humanity faces. Evidently in a book like this we can only consider a selection of issues under each of these headings, but we have chosen topics which we judge to be centrally representative of the demands for climate honesty now confronting us. The overview which follows is to assist readers in picking their own preferred path through the material, which is not intended to be read in any particular sequence— though we would recommend beginning in any case with the Introduction, an argument in more detail for the starting-point sketched above.

Under the rubric of *Politics*, Richard McNeill Douglas's opening chapter asks whether the current economic system, capitalism, as presently configured builds in kinds of imperative which make serious climate mitigation impossible—and considers some implications of the (unsurprising) answer. This is followed by Peter Newell's chapter exploring how a global temperature rise of 4°C is likely to impact the international political order, especially its security aspects. Shifting the perspective, Rupert Read and Kristen Steele then argue that coming climate disasters need not result in a doubling-down of liberal-capitalist non-solutions, but can create opportunities to strengthen or rebuild local polities and economies.

In terms of the *Systems* through which policies take effect, representative issues for land, water and the atmosphere are addressed in turn. Jonathan Essex considers how the trend towards further carbon-intensive urbanisation could be halted with a transformative model for the design and planning of cities. As a case-study of likely specific effects of climate change on a post-growth Britain, Anne Chapman focusses on a 'small-scale' climate disaster, the serious flooding in Lancaster in the winter of 2015–16. At the global scale, Helena Paul and Rupert Read call out the dubious imperatives driving us towards hubristic attempts to geoengineer the climate, reasserting the vital role of the precautionary principle in response.

Crucial to how we tackle these and related questions of politics and systems are the *Framings* of the wider issues within which these questions lodge: the assumptions that go unchallenged in our present understanding of these issues, how we might foreground these assumptions for reflective reconsideration, and how we might come thereby to think differently about our prospects. One important mode of such foregrounding is through historical triangulation, and Brian Heatley's contribution offers a perhaps unexpected comparison of our present situation with an earlier human experience of (albeit rather milder) climate change, that associated with the Crisis of the Late Middle Ages in Western Europe.

Thinking honestly about climate reality nevertheless hurts, and people very understandably flinch from it. Nadine Andrews and Paul Hoggett of the Climate Psychology Alliance explore the psychological and psycho-social dimensions of this resistance, and how it might be overcome.

While hope seems indispensable for coming through climate disaster and still avoiding catastrophe, staying hopeful might seem the most difficult framing of all if we genuinely aspire to climate honesty. My own chapter concluding this section suggests how, suitably guarded against utopianism, hope might remain realistic even against apparently overwhelming odds.

The whole book is about looking to the future, but the concluding *Coda* (written on behalf of the Green House collective as a whole) asks explicitly: where do we go from here? Where for honest thinking about climate disaster, where for those individuals and groups who commit to such thinking, where for any society which at last, as disaster starts to strike, starts to listen—and where, out of all this, for an honest politics? While not yet in a position to offer answers to these questions—and *nobody* is in any such position—we suggest a constructive approach to taking them forward.

We fully expect, of course, that this approach in itself, and its presuppositions, will provoke disagreement, debate and further questioning. To get that dialogue onto the terrain of climate honesty is really the point of the book.

NOTES ON CONTRIBUTORS

Nadine Andrews is a visiting researcher at the Pentland Centre for Sustainability in Business Lancaster University. Prior to this she worked in the science team of the Intergovernmental Panel on Climate Change Working Group II technical support unit, focussing on psychology, social sciences and ethics. Nadine uses mindfulness and nature-based approaches in her independent practice as a coach, consultant, researcher and trainer, and specialises in integrating ecopsychology with ecolinguistics. She is an executive board member of the Climate Psychology Alliance UK, a steering group member of the International Ecolinguistics Association, and a member of the Association for Psychosocial Studies.

Anne Chapman was an environmental consultant for eight years before doing a masters then PhD in environmental philosophy. She is the author of *Democratizing Technology; Risk, Responsibility and the Regulation of Chemicals*, Earthscan (2007). Between 2003 and 2011 she was a Green Party councillor on Lancaster City Council. She is a director of Green House.

Richard McNeill Douglas is a PhD candidate based at Goldsmiths, University of London funded by the Centre for the Understanding of Sustainable Prosperity (CUSP). His research on contemporary attachment to the idea of indefinite growth is contributing to CUSP's work on 'meanings and moral framings of the good life'. Previously he worked as a committee specialist at the House of Commons Environmental Audit Committee. He has written extensively on the environment, economics, and politics in a wide range of publications.

Jonathan Essex is a chartered engineer and environmentalist. He has worked for engineering consultants and contractors in the UK, Bangladesh and Vietnam. This work has included developing strategies and business plans for reuse and recycling, and for decarbonising the UK construction and housing industries. His current work focuses on improving the sustainability and resilience of livelihoods and infrastructure investments worldwide. He also serves as a councillor in Surrey.

John Foster is a freelance philosophy teacher and an associate lecturer at Lancaster University. He is the author of *The Sustainability Mirage* (2008) and *After Sustainability: Denial, Hope, Retrieval* (2015), and has also edited several collections on relevant themes.

Brian Heatley is a former senior civil servant and was co-author of the England and Wales Green Party's General Election Manifestos in 2010 and 2015. He also studied history at Sheffield University in the 1990s.

Paul Hoggett is Emeritus Professor of Social Policy at UWE, Bristol where he was co-founder of the transdisciplinary Centre for Psycho-Social Studies. He is a psychoanalytic psychotherapist and was founding editor of the journal Organisational and Social Dynamics, a forum for those working within the Tavistock Group Relations tradition. He is a Fellow of OPUS, an Organisation for the Promotion of Understanding of Society. He was co-founder and first Chair of the Climate Psychology Alliance which seeks to bring insights from depth psychologies to our understanding of collective paralysis in the face of dangerous climate change.

Peter Newell is Professor of International Relations at the University of Sussex. In recent years his research has mainly focussed on the political economy of low carbon energy transitions. Besides working for academic institutions including the universities of Sussex, Oxford, Warwick and East Anglia, he has worked for Friends of the Earth and Climate Network Europe. He sits on the board of directors of Greenpeace UK and is a board member of the Brussels-based NGO Carbon Market Watch. He is associate editor of *Global Environmental Politics* and sits on the board of several other journals. He is the author of *Globalization and the Environment: Capitalism, Ecology and Power* (Polity, 2012) and many other books and articles

Helena Paul is co-director of EcoNexus, founded in 2000. She has worked on indigenous peoples' rights, tropical forests, oil exploitation in the tropics, biodiversity, agriculture and climate change, patents on life, genetic engineering (GE) and a range of other issues. As well as numerous reports, papers and briefings, she co-wrote: *Hungry Corporations: Transnational Biotech Companies Colonise the Food Chain*, published by Zed Books (2003).

Rupert Read is a Reader in Philosophy at the University of East Anglia, and a specialist in the philosophy of Wittgenstein. His books include *Kuhn* (Polity, 2002), *Philosophy for Life* (Continuum, 2007), *There Is No Such Thing as a*

Social Science (Ashgate, 2008) and *Wittgenstein among the Sciences* (Ashgate, 2012). His other key research interests are in environmental philosophy and philosophy of film. His research in environmental ethics and economics has included publications on problems of 'natural capital' valuations of nature, as well as pioneering work on the Precautionary Principle. He is chair of Green House, and a former Green Party of England and Wales councillor, spokesperson, European parliamentary candidate and national parliamentary candidate.

Kristen Steele has worked for Local Futures/International Society for Ecology and Culture since 2000 and is currently Associate Programmes Director. She co-coordinates the *Economics of Happiness* and *Global to Local* programmes, including the International Alliance for Localisation (IAL). She holds a BA in Environmental Studies and a Master's degree in Wild Animal Biology from the Zoological Society of London (ZSL) and Royal Veterinary College, University of London. She writes regularly for both alternative media and academic journals on topics including environmental crises, economic globalisation and localisation, conservation biology, and wildlife economics.

INTRODUCTION: LOOKING FOR HOPE BETWEEN DISASTER AND CATASTROPHE

Brian Heatley, Rupert Read and John Foster

At the advent of danger there are always two voices that speak with equal force in the human heart: one very reasonably invites a man to consider the nature of the peril and the means of escaping it; the other, with a still greater show of reason, argues that it is too depressing and painful to think of the danger since it is not in man's power to foresee everything and avert the general march of events, and it is better therefore to shut one's eyes to the disagreeable until it actually comes, and to think instead of what is pleasant. When a man is alone he generally listens to the first voice; in the company of his fellow-men, to the second.
— Tolstoy in *War and Peace* (1849, 886), on the consequences for Russia of the French invasion of 1812

Why isn't more being done about dangerous human-triggered climate change? Why isn't the world responding adequately to all the increasingly-dire warnings from climate scientists and from advocates of the Precautionary Principle? Why do we find it as hard as we evidently do to think about a world where the climate has already changed massively—so that we veer between 'it won't make much difference, everything is going to be fine' and 'it's the apocalypse, the end of the world, there's nothing we can do', while refusing to explore the awful, but more middling realities?

Familiarly enough, possible explanations for this mental paralysis include concern that responding adequately to human-induced climate change would make current patterns of economic growth and 'development' unravel—and that 'we can't let that happen'. There is also a mismatch between the climate crisis and the political institutions and ideologies being looked to for solutions, especially under neoliberalism: above all, the rising star of 'freedom' and the (correct) perception that libertarian-style freedom for capital is incompatible with the kind of concerted action that would make climate sanity and climate safety possible. This is only emphasised by the deceptive, self-interested role played by the fossil fuel and various other high-emissions industries. The election of Trump put a climate-change denier in charge (at least notionally) of the nation which has not only produced more emissions historically than any other but is still the second-biggest emitter,[1] and this apparent legitimation of dangerous irresponsibility is already hindering international efforts to address the issues, as was very evident at the 2018 United Nations (UN) Framework Convention on Climate Change conference in Poland.

Beyond these factors, there is also a pervasive feeling that dangerous climate change remains remote, abstract and diffuse, and lies in the relatively far off future. In fact, of course, significantly damaging climate change is already happening, and seriously impacting various low-lying and otherwise vulnerable regions of the world. The level of persistent greenhouse gases currently in the atmosphere already guarantees that this damage will worsen for a long time to come. The effects thus far of those greenhouse gases already in the atmosphere continue to exceed the worst expectations of modellers. Last year (2018) was the fourth warmest on record, according to the European Union's Copernicus Climate Change Service, resulting in California and Greece suffering severe wildfires, Kerala in India having the worst flooding since the 1920s and heat-waves striking from Australia to North Africa.[2] Climate scientists anticipate at least a 2°C increase by the end of the century, and many believe that this is in fact a conservative projection, with the increase likely to be at least in the 3–4°C range and possibly (especially if recent trends extrapolate) much higher. Indeed the continuing exploitation of fossil fuel deposits, and especially the massive increase in hydraulic fracking anticipated over the next fifteen years, suggests that the latter figure is probably more accurate, as a likely *minimum*. But policies promoting economic growth remain hegemonic across the political mainstream. Serious doubts surround the efficacy of so-called 'green growth'. For the projected increases in global aviation and in intensively-reared industrial meat, renewable energy is not an option and eco-efficiency has limited potential. And there is very little prospect indeed of any of this changing anytime soon.

We want in this book to explore what truly facing up to the reality of this situation means in practice.

---

1 And possibly still the biggest, once one takes proper account of outsourced emissions.
2 See https://globalnews.ca/news/4823684/climate-report-eu-2018/.

Being honest about climate change

Virtually all 'mainstream' treatments of this issue—starting with hugely-influential and deliberately optimistic official exercises like the Stern Review (Stern, 2007) but including even work like that of Naomi Klein (2014) and George Monbiot (2017) which accepts the end of 'growth' and sees the issues much more clearly—pretend in various ways that somehow we could continue to have 'prosperity' (perhaps reconceived as 'green growth') while preventing massive climate-driven damage. They pretend, in a spirit of optimism, that somehow or other our lives could still get better and better if only we could unite to overcome the challenges facing us. Even the most clear-sighted mainstream commentators, that is, urge us to action with what are still essentially *hypothetical* horrors: this, appallingly, is what the world to which we have got so comfortably accustomed will turn into, *unless...*

But the stark, categorical truth is that things are now certainly going to get worse—much worse—*whatever* we do. Hardly anyone wants to admit this. It sometimes seems as if almost the entire world is engaged in climate-denial—not just those with vested interests in minimising the dangers, but also those working passionately to save us from them. This is nonetheless perfectly understandable, because what is now unavoidably coming can (and should) fairly be described as *disaster*, or as a connected set of disasters. It will include all the consequences of a rise in global atmospheric temperature which is already locked in, and is on track to increase to 4°C above pre-industrial levels—the Arctic ice-cap gone, sea-levels risen by several metres, large swathes of the tropics uninhabitable or agriculturally useless, and corresponding geopolitical turmoil beside which our present challenges regarding migration will look like a picnic. It will remove the option of a secure and materially abundant Western lifestyle not only from those who have long been incited to aspire to it, but (even more perilously) from those who currently enjoy it. There is no way that all this won't be comprehensively disastrous, whether we think in terms of human material well-being or of 'rights' to security, equality, advancing freedom and other long-established human life-expectations (or, of course, of the effects on other species and on the biosphere itself). As our epigraph from Tolstoy wryly emphasises, screening out approaching disaster until well past the eleventh hour is a deep-seated characteristic of collective humanity. And that applies not just to self-interested denial that it is our actions which are causing disaster, but to optimistic denial that disaster is what they must cause.

In response, this book seeks to manifest *climate honesty*. It considers why we refuse to face the reality of our situation or think straight about how that reality will unfold. It explores various aspects of the programme of *transformative adaptation*[3] which might enable us, collectively, to do something as yet hardly attempted: to take seriously that we are now as a species and a planet already committed to some degree of escalating climate chaos, so that the state of the world is going to worsen significantly for some considerable time to come, but *yet*, at the same time, not to give up.

Very few current voices fully accept the coming disaster of a deteriorating climate while holding onto the possibility of transformative adaptation to save ourselves (if we are lucky) from conclusive catastrophe. We seek to build on what little there is—on ideas such as those of George Marshall (2014) and the Climate Psychology Alliance (http://www.climatepsychologyalliance.org/) in relation to the psychology of the situation, of *Climate Code Red* (http://www.climatecodered.org) in relation to public policy, and on critiques of the counter-productivity of 'sustainable development' narratives in the current context, for example Foster (2015) and 'Dark Mountain' (http://dark-mountain.net). We aim to add perspectives from philosophy, political economy, history, practical experience, the arts and more. Our aim is to pursue climate honesty rigorously and deeply without counselling despair. We will be deluded neither by optimism of the intellect nor by pessimism of the will.

That is what makes this a pioneering book in the literally vital but now increasingly crowded field of climate change thinking.

Recognising the inevitable

But the book is pioneering in large part because, as already noted, two of our principal starting assumptions—that dangerous climate change is now inevitable and that we are going to have to live in a post-growth world—are very far from generally accepted. Since these are for us givens, and as such are not explicitly contended for anywhere in the body of the book, we will summarise the arguments for them here.

---

3 In this book, we use and illustrate this vital concept of 'transformative adaptation' in a way which has similarities to the IPCC's concept of 'transformational adaptation' (https://www.weadapt.org/knowledge-base/transforming-governance/transformational-adaptation), but goes significantly further. For us, transformational adaptation is not only mitigative, but also changes our society in the direction that we need to move in anyway: towards ongoing fleet-footed flexibility in the face of the continuing damage that is coming our way in the 21st century, and towards what human and non-human flourishing remains possible through this damage. Crucially, our assumption is that genuinely transformative adaptation requires giving up on fantasies of endless 'development' (contrast here https://www.wri.org/our-work/project/transformative-adaptation)—on this, see also the discussion in the section on hope, below.

Our first assumption is that it is unreasonable to suppose that dangerous climate change involving a temperature rise of at least 4°C by 2100 can now be avoided, despite the Paris Agreement with its talk of 2°C and subsequent iterations of the UN process out of which this agreement emerged. Quite independently of Trump and Bolsanaro and of the way in which other big players like China, Russia and Saudi Arabia react to a potential US-prompted free-for-all, quite irrespective of whether the UK post-Brexit will abide by EU targets or of other similar unpredictabilities, the Climate Agreement signed in Paris at the end of 2015 would not anyway have saved the world.

The headlines said that at Paris the world's nations had agreed to limit climate change to 2°C, and even recorded an aspiration to keep it within 1.5°C. Climate scientists have subsequently reached a consensus that even 1.5°C is already too dangerous (IPCC, 2018)—in any case, it is plainly impossible to achieve. Global average temperatures were already 1°C above pre-industrial levels at the time of the Paris meeting (*Guardian* 09/11/15), and further warming to 1.5°C would now be very likely indeed even if, *per impossibile,* no further greenhouse gases at all were emitted (King 2016). A global rise of only 2°C entails substantial harmful climate change, and even the total inundation of some low-lying island countries. But *Paris will not contain the rise to 2°C,* even were Trump, unimaginably, to turn into a fervent environmentalist.

To see this we have to see what the substance of the Paris Agreement actually was. Paris amounts to a series of unilateral 'Intended Nationally Determined Contributions' (INDCs) by individual countries. These are, of course, just voluntary undertakings. The EU countries (including Britain) for example have promised collectively a 40% reduction in domestic greenhouse gas emissions by 2030. China says that its emissions will peak in 2030 at the latest, and that it will lower the carbon *intensity* of GDP by 60–65% below 2005 levels by 2030. The US had undertaken, before its withdrawal from the agreement, to reduce net greenhouse gas emissions by 26–28% below 2005 levels in 2025. And so on. In total 185 countries covering around 94% of world emissions made such promises (ClimateActionTracker 2015).

The small effect of Paris can be seen in the graph below. The line on the left shows what has happened up until 2010. The dotted line from 2010 to 2030 shows what was expected to happen without Paris. The still rising dashed line from 2010 to 2030 shows what is expected if Paris were fully implemented. No, it's not very different.

To demonstrate that the 2°C target is now impossible, we can use the IPCC's estimate that to stay within 2°C no more than 1000 gigatonnes of CO_2 equivalent could be emitted after 2010 (IPCC 2014, 10). We now add to that stock at a rate of about 50 gigatonnes a year (see graph). Even with the Paris pledges we will go on emitting over 50 gigatonnes a year for the next 20 years, or a total of 50 times 20 which equals 1000 gigatonnes. It defies everything we know about the longevity of energy infrastructure investments and how the economy works to suppose that emissions will just stop altogether in the 2040s, especially in growing economies.

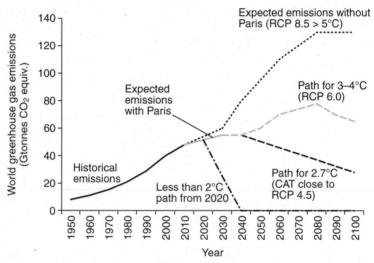

Figure 1. Paris will not make much difference: historical and projected world greenhouse gas emissions 1950–2100. (*Source*: authors' construction based on (UNCC 2015, 11) combined with (IPCC 2014).)

So, barring a miracle, 2°C must inevitably be substantially breached. Nor has anything which has happened since 2015 suggested any reason for doubting that judgement. In particular, the 24th Conference of the Parties to the UN Framework Convention on Climate Change, mentioned above, which spent so much expensive international effort on settling the 'rulebook' for putting the Paris agreement into practice (specifying how governments will measure, report on and verify their emissions-cutting efforts),[4] can be seen as achieving no more than an elaborate seating-plan for the sun-deck of the Titanic.

A key deadline within the Paris process is 2020, when countries are supposed to show that they have met emissions-cutting targets to date, and to

---

4 See https://www.theguardian.com/environment/2018/dec/16/what-was-agreed-at-cop 24-in-poland-and-why-did-it-take-so-long.

affirm new, much tougher targets. But even if a radical programme of reduced emissions was started at that point, and one that went far beyond the Paris Agreement, it would need to achieve zero emissions by 2040 to stay within the 2°C limit (the plunging line on the graph). This is because by 2020 there will have been a further 10 years of emissions after 2010 at 50 gigatonnes per year making 500 gigatonnes. So after 2020 there are only 500 gigatonnes left available in the carbon budget. If we were to reduce to zero emissions by 2040, we should have 20 years starting at 50 gigatonnes a year in 2020 and required to reduce to zero in 2040). On any realistic estimate, this rate of reduction is simply not going to happen.

So what *will* happen? Surely the most optimistic assumption we are entitled to make based on current political agreements and actions across the world is that emissions will continue to rise after 2030, hopefully levelling off later in the century—though this is more hope than experience to say the least. This is broadly represented by the curve on the graph associated with a 3–4°C rise, and is endorsed by a UN Environment Programme report which estimates that we are actually on track for global warming of up to 3.4°C on the basis of the Paris Agreement being implemented (UNEP 2016, p. xi). One of the authors of this Introduction actually predicted this as the Paris Agreement was being reached in late 2015 (Heatley, 2015). More *realistically*, we might instead expect the future simply to reflect the past, and follow the top curve, which is the IPCC's 'Business as Usual' case, where the temperature rise is in the range 4–5°C.

We must assume, therefore, that the level of global temperature rise associated with carrying on as we are, limiting emissions to broadly current levels—that is, at the very least a 3–4°C rise by 2100, and more likely a 4–5°C—will now happen. This estimate is *strongly biased on the side of optimism*:

- it is based on cautious, consensus IPCC estimates. Many individual experts are far more pessimistic, or are more pessimistic about the effects of a particular temperature rise, for example on sea level rise;
- it depends on countries sticking to Paris. Two years after Trump's election, and with an uncertain future for the EU, this looks increasingly improbable;
- it assumes that after 2040 emissions will begin to take a downward path, although—even with Paris—emissions will have risen steadily until then;
- it only takes us to 2100, and the world will continue to get hotter after that, even if there were no further emissions (the deep ocean may take centuries to fully respond to a warmer earth); and

- it takes no account of highly-likely feedback effects, such as the consequences of the release of methane trapped in the tundra or beneath the oceans caused by warmer temperatures, or the loss of carbon to the atmosphere now stored in peat and soils due to lower rainfall, or the fact that as Arctic ice melts the resulting ocean absorbs more heat rather than reflecting it back.

Finally (a point often overlooked) it assumes that our climate models are accurate—yet they have only been tested in a relatively narrow range. Any complex non-linear situation of this type can throw up surprises, including sudden very rapid changes, and changes in the opposite direction to that expected. We must expect 'unknown unknowns', and under such conditions of extreme uncertainty be prepared to take exceptional measures to avoid the worst case. Nevertheless we could accept a possibly modest 4°C estimate without undermining the central and we think irrefutable conclusions that climate change of at the very least this order will now inevitably occur (unless something very fortunate and very unlikely occurs, such as the kicking in of some unknown natural 'negative feedback' mechanism dampening down the temperature increase),[5] and that we need to get used to the idea. It is simply too late for the temperature rise to be any less.

Economic growth will not save us

Our second starting assumption is that we must henceforth live in a post-growth world. For full detail on this, we must refer the reader to Green House's earlier book already cited (Blewitt and Cunningham, 2014). In outline: further net growth, at least in countries like the UK, is neither *necessary*, *desirable*, nor *possible*.

Such further growth is not *necessary* both because humanity has enough, if only we distribute it fairly, and because we can run our affairs in a way that doesn't depend on continual growth, if we are willing (for instance) to share around the work that there is. Our earlier book sets out in some detail how a post-growth world, and especially a post-growth Britain or similar 'developed' country could work—see also Tim Jackson (2009) and Peter Victor (2008).

---

5 Some readers may wonder why at this point we don't mention the possibility of a climate-engineering 'technofix' that could allegedly head off catastrophe by keeping us below 3–4 degrees of over-heat. The reason we don't consider such an alleged possibility viable is explained in detail in the chapter below by Helena Paul and Rupert Read, which thoroughly critiques such 'geoengineering'.

It is not *desirable* because the dirty secret of growth is that it excuses inequality while everyone is promised jam tomorrow (Guardian 2015), and because growth is increasingly a matter now of a dangerously destabilising financialisation. And finally, it is not *possible* because we are breaching the planet's boundaries—or rather, it is possible now only at the cost of being much more clearly undesirable through the accumulating consequences of transgressing those limits.

Economic growth should be understood from the point of view of the long term of hundreds of years, the '*longue durées*' of the French Annales historians, as at best the transitional stage between one steady state and another. Either that second steady-state is at hand, or we are going to face an uncontrolled collapse: at least an economic collapse, probably some kind of civilisational collapse, conceivably a complete collapse of complex life on Earth. These outcomes would probably produce a new steady-state in due course, one far less hospitable to human life as we currently know it. Climate catastrophe is only one possible mode in which such collapses might be engendered. But it is now, as everything in this book corroborates, an alarmingly likely one.

'Green growth' is widely touted as the answer to the problem just set out. It is a fallacious answer, as we showed in detail in our earlier book (see especially the summary at 'Why not just embrace "green growth"?' (Blewitt and Cunningham 2014,183)). In brief: there is no successful historical instance *nor even any successful economic model* of *net* green growth (as opposed to simply green growth in specific economic sectors, such as renewables, which is of course something to be aimed for and applauded). But it is *only* with net green growth that we could hope to meet even the kind of climate targets aimed for in the Paris Agreement, let alone sounder targets. Moreover, as should be obvious, setting growth targets of any kind is simply bound to make the task of hitting climate targets harder. For every additional scintilla of growth, one needs to make the carbon de-intensification of economic activity that bit deeper, and so that bit harder to attain.

Assuming that growth has come to an end has a devastating effect on one of the main arguments for effectively ignoring climate change, or at least doing nothing to mitigate it. For some economists, the question of what to do about climate change is largely a matter of comparing the costs of mitigation with the costs of adaptation. If mitigation costs *more* than incremental adaptation, then don't bother to mitigate (Lomborg 2001, Lawson 2008). However, this compares costs separated by a long period of time; we have to mitigate now, while our successors will be adapting tens and hundreds of years later. If one assumes extensive economic growth in the interim, then the proportion of national income taken up by even quite extensive costs of much later adaptation is considerably reduced in comparison to the costs of mitigation now, measured the same way—because the denominator, the national income, is *assumed* to have grown massively in the intervening period. Put more simply: because of continued growth, our successors will be so much richer than us that they will have to expend relatively little effort in clearing up the mess. This argument—which never worked morally, since it has quite different people doing the damage and bearing the costs—doesn't work even *arithmetically* without assuming substantial growth.

Economic growth will not save us. We come most certainly not to praise it, but to bury it. Otherwise, it will bury us.

The future will be post-growth. This will come about either by design (as we would prefer), or by accident (as 'nature' forces upon us all a massive 'correction'). But here we reach a dilemma. For this, our second assumption, while gradually gaining in intellectual respectability as it becomes clearer and clearer to the open-eyed how our mainstream economics is pushing us toward catastrophe, is nevertheless still not only not widely shared; worse, it is normally openly rejected, or simply 'framed out': not even allowed to enter into consideration. As we mentioned earlier, the hard truth is that one key reason why collectively we are unwilling to face up to the reality of man-made dangerous climate change is precisely that doing so requires us to face up to the need for a shared and urgent post-growth project. Indeed, some have argued persuasively that it is not just a matter of post-growth, but that the wealthier nations have to embrace de-growth, or contraction (Anderson 2013). People refuse to face up to climate reality—almost everyone is in 'soft' climate denial (as opposed to the 'hard' climate denial of Trump and those who ape him)—because they worry that it will make growth impossible, and so will cause the story of unending human material progress to unravel.

This places us in a difficult position. The way we propose to address this difficulty is by trying to be honest about our species' shared predicament, willing readers to join with us in accepting both our key assumptions. Like it or not, with the climate on the trajectory which it can't now avoid, we are living in what is fundamentally, one way or another, a post-growth world.

But we are the first to admit the very arduous nature of this task. It is part of why facing up to climate reality is so necessary. As a species, we are *not* going to embrace post-growthism in good time—which would be, basically, *now* (or, better still: two or three decades ago...). It is going to take a long while to turn the global political-economic supertanker around and head it towards a saner alternative. We can only go on in the hope that this will happen prior to runaway climate change taking hold—the most likely route, so far as one can presently tell, by means of which the Earth may cause societal collapse and bring on final catastrophe. A wonderful feature of human beings is that we have *in theory* the capacity to engage in extremely rapid change, once we realise that there is a serious problem at hand. The challenge is to engage this capacity in practice.

Thinking about these hard consequences of an average global temperature rise of 4°C means thinking about the future, and the future well beyond next week or even the next several years, which is the 'medium term' to which current economic policy in particular confines itself. As noted above, we need to think in terms of the '*longue durée*', a span of maybe a hundred years or more. Thinking about the future in this way is also hard in another sense, as shown by so many attempts to do so in the past having turned out to be clearly and demonstrably wrong. We are largely not, as Keynes thought in the 1930s that we might be, living in a future where what to do with our vast leisure time is our principal preoccupation—instead, many of us are working longer and more frantically than ever. Nor are we all buzzing around in personal aircraft living off nutrition tablets as some of the science fantasists would have had us do. Nor has the rate of profit fallen so low that the inevitable final crisis of capitalism has led to socialist revolution, as confidently prophesied by Marx. So one embarks on talking about what might happen in the rest of this century with some humility and considerable apprehension. Most people avoid it altogether, although (worryingly) those who are busy thinking about the future include, as Peter Newell points out in his contribution, the military authorities of the leading powers and some corporations, behind closed doors. If it were not so certain that the world faces disastrous change, it might be entirely reasonable to adopt as the default assumption the usual one, that the future, in most respects, will be very much like the past, only improved on by steady but constant progress. That is how most of us conduct our daily lives, and in the short to medium term it has worked very well.

But we do now, and inevitably, face climate change bringing disasters in its train, in a world where economic growth, at least in the rich world, is halted. We must surely begin at last to face up to this.

Very probably a lot of our predictions and expectations will be wrong. But if we assume instead that the future will be the past plus progress, we can know *for certain* that we will be wrong.

And hope?

The last two sections have been about destroying illusions, a painful but also a necessary process. What room do they leave for hope? The answer, we take it, depends on how robust the suggestions for transformative adaptation contained in the following chapters are found to be.

Adaptation is transformative when it is premised not on preserving as much as possible of our present world—of uninterrupted human 'progress' and material prosperity—as is compatible with accommodating the climate and ecological damage which we have already done, but rather on responding to that damage through *open-ended* change in our institutions, practices and policies across the board. Transformative change is open-ended precisely because it trans-*forms*, that is, the patterns defining our field of activity are recast, so that in the nature of the case we cannot know beforehand where we will end up. Such transformation requires, above all, readiness to *let go* of the form of life for which 'progress' and material prosperity are unquestionable givens. Inevitably, therefore, it has a tragic as well as a creative dimension, and we must be prepared for both (Foster 2017a, 2017b).

The kinds of transformative change canvassed in the various chapters of the book include moves towards a post-capitalist economy, a significantly reconfigured world order, greatly increased localisation of social and economic decision-making, carbon-intelligent land-use planning, much greater community resilience and fundamental acceptance of the precautionary principle as a policy guide at all levels. These all depend, in turn, upon transformations in our ways of thinking—in our attitudes towards what we can learn from the past, towards our own deeply-embedded attitudes and towards the conditions for being realistically sanguine about our transformational prospects themselves.

This is a daunting agenda, but it is entailed on us by any real commitment to climate honesty. To the extent that even some of it is persuasive, we would contend that there is genuine hope to be found between the honest recognition of oncoming disaster and the final human catastrophe which remains, even yet, possible to avert. We hope that our book is catching—and strengthening—a moment at which humanity summons enough courage to look our morbid present and our probably-awful future in the face. Only by doing so, and recognising that, while it is too late for mitigation alone, a programme of transformative adaptation is now as viable as it is urgent, can we rise up to meet our real situation. It is in that spirit that we aspire to change minds, and thus to influence the actions which might change that reality.

Part I: Politics

Chapter 1

COULD CAPITALISM SURVIVE THE TRANSITION TO A POST-GROWTH ECONOMY?

Richard McNeill Douglas

> For some people, growth and capitalism go together. Growth is functional for capitalism. Capitalism demands growth. The idea of doing without growth is tantamount to doing away with capitalism, in this view.
>
> — Jackson (2017, 222)

One of environmentalism's key tenets is a belief in the limits to growth, the thesis that 'infinite growth is impossible on a finite planet' (Connelly and Smith 2002, 52). As is well known, this has often set green thinkers in conflict with mainstream (neoclassical) economists, who tend to treat an abstract idea of the economy as being capable of indefinite expansion. What is less often discussed are the divisions on a different question among those heterodox economists who *agree* on the limits to growth thesis: assuming that the pursuit of limitless growth must be abandoned, must this also require the simultaneous abandonment of capitalism?

On one side of this debate are those who combine environmentalism and Marxist political economy. For these 'eco-Marxists',[1] the limits to growth

---

1 While the term 'ecosocialist' is perhaps more familiar, I am preferring not to use it as it is often associated with the journal *Capitalism Nature Socialism* and past editors such as James O'Connor. This group has had sharp differences with another circle of Marxist environmentalists, associated with the journal *Monthly Review*, including long-time editor John Bellamy Foster. Notwithstanding the enmity within these disputes, for my present purposes I am seeking to treat them all together. For discussion of these divisions, see Proyect (2016).

thesis is joined together with one of the key tenets of Marxism, that capitalism requires endless accumulation to operate, that it must 'grow or die'. For eco-Marxists any form of post-growth economy cannot by definition be a capitalist one. Against this position stands a school of ecological economists—often associated with Herman Daly's concept of 'steady state economics'—who tend to remain agnostic about whether capitalism might in some form survive the transition to a post-growth world.

Of course, there are other schools of thought among environmentalists (e.g. 'light greens') and heterodox economists (e.g. many post-Keynesian economists) who believe that, with sufficient state investment and innovation, growth can be made sustainable for the foreseeable future (resulting in 'green growth'). For some who subscribe to these schools of thought, questioning the nature of a post-growth economy or the survivability of capitalism is an unnecessary distraction. Robert Pollin (2018), for example, makes a compelling argument that the 'climate-stabilization imperative' requires an urgent *growth* in renewable electricity and energy efficiency investments, meaning that talk of an overall reduction in economic activity is at the present moment beside the point.

Against this, however, one could argue that the urgent need to invest in decarbonisation does not change the fact that we live on a finite planet. Besides which, in addition to the ecological imperative of capping our use of resources, there are good reasons for believing that the conditions that have given rise to economic growth are rapidly giving out now, resulting in a future of 'secular stagnation' (Jackson 2018). In such a context, the transition to a new economic reality, one which does not depend on indefinite growth, is a subject which demands to be grappled with theoretically.

To focus on the question as to whether such a transition to a post-growth reality must also mean imagining the end of capitalism, the terms of this debate between those agnostic about capitalism on the one side, and the eco-Marxists on the other, *had* remained much the same for years since the rise to prominence of the limits to growth thesis in the 1970s. Recently, however, new research by Tim Jackson and Peter Victor (2015) among others, on the possibility of eliminating a 'growth imperative' from a market system, has resulted in modelling of an economy which does not grow, does not crash, and yet retains many of our currently-recognisable forms of capitalist organisation. Might this signal a theoretical victory for the agnostics in this debate?

Perhaps not; perhaps in fact it is the beginning of something new, leading to the dissolving of some of the boundaries between different schools of heterodox economic thought. This is to say, the models developed by the likes of Jackson and Victor still require a variety of special conditions to enable them to work. In this sense, they would still imply a radical transformation of capitalism—even to the extent that it began to blur into socialism. To understand what this means we need, first, to engage with the debate between agnostics and eco-Marxists to date; and second, to rehearse the argument that there is an essential 'growth imperative' within capitalism. After that we will be in a better position to consider how the new developments in sustainable macroeconomics could change this debate.

Contours of the debate

Before reviewing the exchanges in this debate to date, we might want to ask: why does it matter? There are perhaps two main answers. The first concerns the *political reception* of the proposition that the social pursuit of economic growth must come to an end. Where this idea is presented within the framework of Marxist political economy (in arguing that capitalism must also come to an end) it will generally find itself subsumed within a pre-existing field of political debate. In a negative sense this may reinforce a narrative promoted by neoliberal opponents of environmentalism, for whom it is a 'green road to serfdom' (Postrel 1990). More positively, given the growing popular interest in socialism in the aftermath of the 2008 financial crash, this may conversely win more support for the idea. In either case, however, where post-growth economics is presented as being essentially anti-capitalist this inevitably means it carries an amount of ideological baggage which must be reckoned with before its environmentalist case can, as it were, be allowed to speak for itself. The success of Tim Jackson's 'agnostic' work *Prosperity Without Growth* provides an illustrative contrast. Notably, its success in France has been attributed in part to its very agnosticism as to the potential for capitalism to survive in a post-growth economy; this has enabled it to surmount an often acrimonious dividing line on this question within French environmentalism (Semal and Szuba 2017).

A second reason why this debate matters concerns the *content of post-growth economics* itself, its vision of how economic production, distribution, and consumption would be organised. In part, this is a matter of visualising the texture of everyday life—the nature of work, the kinds of jobs, the roles of workers and employers, the relationship between state and producers, the size and scope of markets, and so on. In part, it is also a matter of modelling under what conditions a non-growing economy could be macroeconomically stable, reproducing its own material basis indefinitely without crashing. It is a question, then, both of its viability as an economic system, and of the political qualities of economic life, particularly relating to individual liberty and material equality. One way of reflecting on this would be to ask whether, if a non-growth economy by definition could never be capitalist, environmentalists need to be arguing for the institution of state socialism and the abolition of all private enterprise? If not, the question remains: what other kind of economic system is conceivable for the provisioning of highly populous, diverse, modern nations—what Friedrich von Hayek (1976) called 'the Great Society'?[2]

The rival positions taken in this debate thus present different orientations towards the problem in hand and the tasks required to address it. Whether or not capitalism could be made environmentally sustainable dictates the character of the political task—in terms of shifting hegemonic ideas and mobilising broad-based movements—required this side of a transition to a post-growth economy. It also defines the theoretical project of designing a successful post-growth macroeconomics, in turn informing the policy tools required to bridge existing capitalist economies through transition into a non-growth form in a managed process.

The debate on these issues is a live one, developing both in content and prominence. We could see this as being due, on the one hand, to a greater sense of urgency, given widespread economic turbulence since 2008 and ever-starker indicators of dangerous climate change; and on the other to growing activity within the still young field of sustainable macroeconomics. As an illustration of this debate in action, it might be useful briefly to focus on the interactions of Tim Jackson, Herman Daly, and Philip Lawn, on the one hand, and eco-Marxists Richard Smith, and John Bellamy Foster and Fred Magdoff, on the other.

In *Prosperity Without Growth*, Jackson begins with a definition of capitalism which makes the question of a growth imperative somewhat tangential. Following Baumol (Baumol, Litan, and Schramm 2007), he defines it as 'where ownership and control of the means of production lies in private hands, rather than with the state.' He then speculates that a post-growth world would see a mixture of ownership models, with plenty of worker-run co-operatives, and a high degree of state intervention, all of which would tend to render classification of the economy a moot point. 'Perhaps', he suggests, 'we could agree to coin the terminology of a 'post-growth capitalism'.' But then: 'Perhaps not.' Ultimately, he suggests it is the 'post-growth' element which is all important; whether such a system need be called capitalism is a 'rhetorical oversimplification' (Jackson 2017, 222, 224).

---

2 Without endorsing his political economy, it seems apt to borrow Hayek's term given the similar challenge he posed to socialists to account for how they would meet the needs of complex societies, and do so moreover in a way which preserved their complexity.

Another leading thinker who, like Jackson, could be considered an 'agnostic' as to the prospects of a sustainable capitalism is Herman Daly. Daly's (1991, 70) proposals for a 'steady-state economy' are based on 'impeccably respectable conservative institutions: private property and the free market'—but he is equally clear that: 'Insofar as capitalism has to grow then it is incompatible with the steady state.' Daly's ambiguity on this question is summed up well by his remarks in a 2018 interview with the writer Benjamin Kunkel. Having said that, 'I wouldn't take the view that we should just abandon capitalism and opt for eco-socialism', Daly goes on to say: 'If you have a Jeffersonian-type, small-scale capitalism, operating within scale and distributive limits, and you want to call that eco-socialism, that's fine with me.' Kunkel subsequently glosses these views as adding up to 'some sort of steady-state social democracy' (Daly and Kunkel 2018, 96, 102).

For the eco-Marxist Richard Smith, approaches such as this are inadequate in that they appear to not to engage with the dynamics of the capitalist system—i.e. the conditions it requires in order to perpetuate itself. Instead, as he describes the agnostics' position, 'growth is seen to be entirely *subjective*, optional, not built into capitalist economies' (Smith 2015, 18). He highlights Bill McKibben's foreword to Jackson's book, where growth is described as a 'spell' which has enchanted us, but which we now need to break. Smith's implied argument is that such language obscures the need for systemic change, suggesting that we can make a difference as individuals simply by changing our attitudes.

One of Smith's main criticisms of steady-staters is their neglect of political economy, understood in its contemporary sense as the meeting of economics and (inherently agonistic) political ideology. While paying handsome tribute to his writing, Smith criticises Daly's steady-state blueprint for society—in which expert bodies would set quotas to limit the use of natural resources, with the market then operating as normal within those limits—as a technocratic vision without any sense of political conflict over the division of a shrinking social wealth. Smith (2015, 24) attempts to bring the debate down to earth: 'Do we need to limit production of meat, coal, oil, synthetic chemicals? How about Starbucks' frappuccinos, SUVs, Flat screen TVs? Ikea kitchens, jet flights to Europe, 12,000 square foot homes? Daly doesn't tell us.'

In his immediate response, Daly (2010, 103) criticised Smith for failing to present a detailed case for how a socialist alternative might function efficiently. 'Instead of markets', Daly asked, 'should we not have another go at centralized rationing of goods and resources, collectivization of agriculture, abolition of exchange and money?'—the implication being that the historical record of such measures has been a dismal failure. If Smith's criticism of the agnostic position is that it is lacking in political economy, then, we might understand Daly's criticism of eco-Marxism as being that it is lacking in economics.

Daly is not the only critic to respond to Smith's arguments. In a notable contribution, Philip Lawn (2011) has suggested that all the features of capitalism that are said to add up to a 'growth imperative' *are*, indeed, essential— but only to the historically-developed institutions of capitalism. It would be possible, he suggests, to redesign capitalism institutionally so that in practice it no longer depended on a quantitative growth in economic production. As with Jackson, he defines capitalism as a matter primarily 'to do with property ownership and the use of markets as a resource allocation mechanism.' This enables him to suggest that an economy could admit a high degree of government intervention and still be classed as capitalist. Growth might then be replaced as the system's animating spirit by a drive for qualitative development which would still (breaking with Jackson here) allow for 'large profits' (Lawn 2011, 2–3).

Lawn's notion of post-growth profit has, in turn, been strongly attacked by the eco-Marxists Fred Magdoff and John Bellamy Foster (2011). Interestingly, Magdoff and Foster (2011, 167) *do* consider the theoretical possibility that capitalism could persist in a 'steady state, no-growth' system, so long as profits are fully consumed by capitalists and workers (i.e. with no accumulation for further investment). However, they reject the actual possibility of such a 'capitalist utopia', on the grounds that:

> it violates the basic motive force of capitalism. What capital strives for—the purpose of its existence—is its own expansion. Why would capitalists, who in every fiber of their beings believe that they have a personal right to business profits, and who are driven by competition to accumulate wealth, simply turn around and spend the economic surplus at their disposal on their own consumption or (less likely still) give it to workers to spend on theirs—rather than seek to expand wealth?
> — Magdoff and Foster (2011, 56)

Here again, we might observe, much of the difference between 'agnostics' and eco-Marxists comes down to the definition of capitalism. Specifically, the issue is the extent to which its static definition as relating to ownership and markets can be separated in practice from the dynamics both of class power and of macroeconomic stability.

Which conception of capitalism is correct? To begin to address this question the next section will look at the case that has been made that there is an essential growth paradigm in 'actually existing capitalism'—the system as

20

it has developed historically, and as we understand it today. Once the main arguments behind this assertion have been unpacked, the rest of the chapter will ask two further questions: first, could this dynamic system be redesigned, in such a way as to lose its growth imperative but still be macroeconomically stable; and, second, could such a system still be described as a form of capitalism?

The case for a capitalist growth imperative

To review the case made for an intrinsic growth imperative in capitalism as we know it today, it makes sense to use—with some additions—the arguments presented by Magdoff and Foster, as leading representatives of an eco-Marxist viewpoint. (In fact, as Magdoff and Foster point out, these arguments are not confined to Marxists, but are frequently the commonplaces of post-Keynesian thought.) We can perhaps distinguish four central though overlapping factors in Magdoff and Foster's presentation, which we shall come to in turn.

The pursuit of profit

The central case made by Magdoff and Foster (2011, 42) is that: 'Capitalism's motive force is the competitive amassing of profits for new capital formation in order to generate more profits and accumulation, *ad infinitum*.' Or, to borrow from another source, James Fulcher's *Capitalism: A Very Short Introduction* (2004, 14), 'the investment of money in order to make a profit [is] the essential feature of capitalism'.

This is indeed the difference between *money* and *capital*: money becomes capital when it is amassed not simply to be spent on the things one desires, nor to be hoarded as savings, but as a fund for ongoing investment, designed to increase itself. 'The object,' Magdoff and Foster tell us (2011, 43), is '[...] to end up with *more money* than one started with, that is, surplus value or profits. [...] Capital, understood in this way, is self-expanding value.'

Of course, advancing money to realise a profit has been the practice of a merchant class since time immemorial—thus capitalists existed long before capitalism proper. What turns the use of capital into the system known as capitalism is when, as Fulcher (2004, 14) writes, 'the whole economy becomes dependent on the investment of capital and this occurs when it is not just trade that is financed in this way but production as well.' Marx famously expressed this dynamic with the schematic M–C–M', whereby money (M) is invested in producing commodities (C), which are in turn sold for more money (M'). As Magdoff and Foster (2011, 43) present the Marxist critique:

'Such an exchange process has no end, but simply goes on and on without limit. Thus in the next round exchange takes the form of M'–C–M", which leads in the round after that to M"–C–M''', and so on in an incessant drive to accumulation at ever higher levels.'

Here we find the central (eco-)Marxist account of a growth imperative as an essential feature of capitalism: that's the entire point of it. There are plenty of charities, and small businesses which are happy simply to tick along and pay themselves a steady wage. But on a large scale and across the economy as a whole, the point of economic production is to make more money than you started out with. Given businesses overall want to do this, then the economy as a whole will keep growing.

The dynamics of competition

If the pursuit of profit is the overarching reason—the *final cause*, we might say—why capitalism is committed to endless growth in this account, there are other essential dynamics in operation—the *efficient causes*—which are said to compel it likewise.

In a classic Marxist account, it is 'because of competition', as Magdoff and Foster (2011, 41) put it, that 'companies are impelled continually to increase sales and to try to gain market share.' This sense of being subject to forces of competition beyond one's control is understood to be an essential feature of capitalism, related to the transition from feudalism to what Polanyi (2001 [1944]) analysed as 'market society'. This is a world in which there is no more important, socially-organising principle than the market. Here, no one is safe from being outflanked, undercut; capitalist society is intrinsically insecure, and in such a situation economic agents react by getting their retaliation in first and continually seeking ways to lower their costs and increase their sales.

As previously suggested, analysis in this vein is far from the preserve of Marxist economists. Marc Lavoie (2015, 123) finds similar analyses of the growth-oriented behaviour of typical firms under the pressures of competition in the views of a range of economists—not only Michal Kalecki, Joan Robinson, and more recent post-Keynesians, but also American Institutionalists such as John Kenneth Galbraith. Having surveyed these economists, Lavoie presents his own post-Keynesian account of how firms respond to the uncertainties engendered by market competition:

> My view is that power is the ultimate objective of the firm: power over its environment, whether it be economic, social or political. [...] [T]he search for power procures security to the individual owner or to the

organization. Firms would like to ensure their long-run survival, the permanence of their own institution.

— Lavoie (2015, 128–129)

The primary response of individual firms to the pressure of competition is thus to do what they can to insulate themselves from its effects. How is this to be achieved? 'The answer is simple: to become powerful, firms must be big; to become big, firms must grow. As a first approximation, it may then be said that if firms attempt to maximize anything, they will try to maximize their rate of growth.' This striving for security, power, and growth can be seen as informing the entire nature of work in the private sector. As Lavoie writes: 'Whereas a successful quest for power will endow the firm with stability and permanence, it will simultaneously endow the individual with a successful career, the opportunity of promotion, the availability of higher social status, respect of peers' (Lavoie 2015, 132, 130). In this way, we can recognise the existence of a growth imperative also at the level of individual employees.

To add a further layer of interpretation, we should remember that one firm's act of striving to insulate itself from the effects of competition manifests itself *as* the effects of competition in respect of the other firms in its market. Competition and the responses to it are one and the same, meaning it—and the orientation towards growth which it impels—is a self-fuelling process. As Joan Robinson (1971, 101) put it: 'Any one, by growing, is threatening the position of others, who retaliate by expanding their own capacity, reducing production costs, changing the design of commodities, or introducing new devices of salesmanship. Thus each has to run to keep up with the rest.'

A final reflection on post-Keynesian and Marxist theories of competition is to recognise that, especially where a handful of firms have grown to the point of attaining oligopolistic status, it is less likely they will strive to compete on price. That is, the mutual recognition that a price war would leave all participants worse off is enough to restrain the atomising effects of the market. In place of price competition, firms will prefer to compete 'primarily through cost-reduction and the sales effort' (Magdoff and Foster 2011, 41).

To focus solely on 'the sales effort' at this juncture, what this translates into is a drive to stimulate desires—most of all, to titillate through novelty, inciting new tastes, new needs, new opportunities to purchase and consume. While each firm is investing in such product development and marketing in order to increase its own sales and market share against its rivals, the collective effect is to expand the market as a whole. This process necessarily leads 'to an explosion in the rates of consumption': in the Western world in the decades since the Second World War, 'We have changed almost every aspect of the way we eat, drink, travel, house ourselves, wash, rest, and play' (Magdoff and Foster 2011, 48).

This 'explosion' in consumption is, in another sense, what Heilbroner (1985, 60) described as 'an implosive aspect' to capitalism, as the market penetrates ever-deeper into the domestic sphere. As he puts it, 'Much of what is called 'growth' in capitalist societies consists in this commodification of life, rather than in the augmentation of unchanged, or even improved, outputs.' We may observe here that this process in itself gives rise to a momentum for growth, in that by drawing more people more deeply into market activity, both as workers and consumers, it makes them less self-reliant. As work crowds out the opportunity and skills needed to maintain people's own households (e.g. cooking, childcare, housework, DIY) and communities (e.g. social activities with friends, finding romantic partners), so they become more dependent on the market to satisfy their needs.

Moreover, if growth is to be maintained, this is a process which *must* continue; businesses must endlessly be looking for new ways to persuade people to give them money. As Lavoie (2015, 135) highlights, in the post-Keynesian account profitability is key to the entire business enterprise: if firms strive for growth in the pursuit of power, then they strive for profit—which facilitates investment—as the means for growth. And this means—given that, 'The growth objectives set by decision-makers are constrained by the financial requirements of profitability, past and expected'—that it is not just past profits but *forecasts of future profits* which are necessary in order to secure the investment required to boost the sales effort and expand production. The system is thus committed to expanding itself: it must keep growing in order to secure its ability to keep growing.

Economic and social stability

While the pursuit of profit and the compulsions of market are the two main factors behind the growth imperative cited by Magdoff and Foster, two others emerge from their account of the dynamics of capitalism. One is the need to offset the destabilising effects which capitalist production exerts upon itself. To return to the analysis of competition taking place 'through cost-reduction and the sales effort', only this time focusing on the cost-reduction, what this translates into are measures which, while good for each firm's competitiveness, are—all things remaining equal—bad for the economy as a whole. Cost reduction typically takes the forms of making workers redundant (through automation, off-shoring, and increasing the workloads of a smaller workforce), repressing wages, and squeezing suppliers. As a whole this means firms spending less money on their workers and each other, and thus receiving less from them; assuming lower spending out of profits by capitalists, the macroeconomic effect is to reduce the flow of demand throughout the economy. The internal dynamics of capitalism thus tend automatically toward a self-generated recession. In this sense, the sales effort and expansion of the market is an essential feature of the system, needed simply to prevent the whole thing from crashing.

These dynamics are certainly affected, it must be acknowledged, by a growth in population. Where population is rising this in itself constitutes a potential increase in the market—an enlargement of consumer demand which can therefore stimulate an increase in employment. At the same time, in adding to the number of workers it adds potentially to the size of the 'reserve army of labour'. In itself, this exercises a downward pressure on wages (and hence effective demand): an excess number of workers helps to keep those in jobs on their toes and in their place, via the implied threat that they can easily be replaced. One need only observe the essential docility of the UK workforce during the years of austerity since 2008, in which time there has been a widespread reduction in real wages, to perceive such effects of job insecurity. Overall, then, a growing population does not in itself sustain a growing economy. As Magdoff and Foster (2011, 59) express it: 'Growth for the economy as a whole—significantly higher than the rate of population increase—is required [...] in order to provide enough jobs to keep unemployment from destabilizing the society.'

Magdoff and Foster's choice of words here, 'destabilizing the society', is apt, in that this is not solely a question of macroeconomic stability, of maintaining effective demand. There is also the question of maintaining popular belief in the system. In this sense, growth has helped to, as Douglas Dowd (1989, 21) puts it, 'camouflage the necessary inequities and inequalities of income, wealth, and power that are intrinsic to the system.' Even if workers' share of the cake has not increased, in other words, growth has enabled them to enjoy a bigger slice in absolute terms, in turn helping to forestall disquiet about their relative share and unequal status.

The role of household debt

The fourth contributor to the growth imperative emerging from Magdoff and Foster's account is an increasing dependence of the system on household debt. The pursuit of cost reduction also leads to an indirect growth imperative, we might say, through the need it imparts to workers to supplement their pay. In the face of stagnant wages since the 1970s, Magdoff and Foster (2011, 55) tell us, 'consumers [have] used their credit cards and borrowed against rising home values to sustain their standards of living', a point reinforced in a UK context recently by Adair Turner (2018).

This same process has been accelerated by the entrance onto the world market of successive waves of low-wage workers in Asia. This phenomenon has helped to lower the cost of products, amplifying the purchasing power of Western workers, and compensating them for wage repression. But at the same time, the competition from low-cost workers has helped to depress Western wages, in turn applying a downward pressure on purchasing power. As Robert Reich (2009, 99) observed in the 2000s, in a specifically American context, 'When we find great deals on cars, refrigerators, picture frames, or almost any other manufactured item, it's often because the Americans who molded, fit, clamped, or bolted these things have either accepted cuts in pay and benefits or lost their jobs altogether.' Growth in private debt has thus been required to offset reductions in effective demand from wages, in turn contributing to an overall growth imperative as a matter of maintaining the ongoing confidence of lenders to carry on lending. The entire consumer economy in this way becomes drawn into a bubble; and the essence of a bubble is that it has to continue to grow—or the effects are calamitous. Magdoff and Foster (2011, 55) link the rise of household debt in response to wage repression to the 2008 crash.

Dependence on interest-bearing debt

The role of bank credit within the dynamics of the capitalist system extends more widely than the rise in household debt in recent decades. While not treated explicitly within the Magdoff and Foster account used here, for many people the role of banks in creating interest-bearing loans is the central plank in their critique of capitalism's intrinsic unsustainability. For this reason, it makes sense to address this separately, as one further element in a widely-made case for the existence of an intrinsic capitalist growth imperative.

This argument is typically articulated as follows: firms require loans in order to finance production; in making loans, banks are essentially creating money 'out of thin air', on which they charge interest as their means of making a profit; this implies that the economy as a whole is in debt, and must therefore grow as a whole in order to repay both loans and interest. An example of such arguments is provided by Philip Goodchild:

> [I]ncreased production and profits require increased investment and debt. [...] Increased production can only lead to the repayment of debts if there is more money to pay for them; yet money can only be created in the form of further loans [...]. The result is that [...] [t]he entire

global economy is driven by a spiral of debt, constrained to seek further profits, and always dependent on future expansion.

— Goodchild (2007, 13–14)

For Richard Douthwaite, a prominent articulator of this argument, all this means 'the economy must grow continuously, if it is not to collapse' (in Jackson and Victor 2015, 32).

This argument as to a capitalist growth imperative is perhaps most strongly associated, not so much with a strict eco-Marxist perspective, as with that of ecological economics or green political economy. The reason for this may be because it depends on a contemporary understanding of banks' issuance of loans, and thus creation of money, essentially *ex nihilo*—in contrast to an earlier understanding of credit, on display within Marx's *Capital*, in which loans are funded entirely from recycled savings. Nevertheless, it is certainly possible to conjoin this more contemporary account of the nature of bank credit with a Marxist analysis of a capitalist growth imperative; much of David Harvey's oeuvre is devoted to this project (see, for example, Harvey [2013]).

Another approach (which I have sought to develop elsewhere) focuses on a contemporary reinterpretation of the thought of Rosa Luxemburg, and her theory of capitalist breakdown (Blackwater 2015). Luxemburg asked a simple question: where does the money come from to enable capitalists to earn a profit and the economy to grow? She made it clear that it could not come from the workers, since when they spend their money on goods and services, capitalists as a whole are only getting back what they have already paid out in wages. The same applies to capitalists' purchases from each other; by definition, as a whole capitalists receive from each other only as much as they spend on each other. What appeared mysterious was where the extra money comes from, to allow capitalists to receive more than they have already paid out, and thus to enable capital accumulation to take place.

Luxemburg's answer was that the extra income had to come from exports to economies that had not yet been brought within the capitalist system. Her breakdown theory was centred on the logical implication that this dependency on a geographical expansionism was economically unsustainable: in order to pay for these goods, colonial economies would end up being converted to capitalism, meaning they would dry up as sources of additional income from outside the system—leading to its inevitable collapse.

Luxemburg's theory has been subject to numerous criticisms, which partly explains why it has remained neglected for so long. Recently, however, her ideas have begun to enjoy a renaissance, thanks to a generation of critics who are drawing on newer understandings of the macroeconomic roles of money and credit—not least money circuit theory, as advanced by the likes of Augusto Graziani (Bellofiore 2013). This representation of the dynamics of a capitalist economy recognises that in capitalist production firms always incur costs (wages, materials, premises, promotion) in advance of sales income, and thus necessarily depend on credit. Given that *as a whole* firms require credit, necessarily they cannot as a whole finance themselves out of retained savings, either amassed individually or loaned to each other; they must obtain their funding from the banks.

A close reading of Luxemburg's theory reveals that she herself recognised that it was bank loans which provided the ultimate answer to the fundamental question she posed. Western manufacturers indeed sought new markets in order to expand their sales, but what provided colonial economies with the effective demand for these exports was itself credit extended to them from Western banks. As she observed in one example, 'this loan capital pays for the orders from Egypt and the interest on one loan is paid out of a new loan' (Luxemburg 1964 [1913], 438).

Under this understanding of Luxemburg's theory, credit is thus what *enables* growth. But it may also give rise to a theoretical account of why capitalism is *compelled* to grow: By providing loans, banks inject new money into the economy. In order for this to lead to an increase in material utility (rather than just leading to an increase in inflation), there needs to be an expansion in the overall value of goods and services that can be exchanged for money. If successive waves of production begin in newly-created credit, then capitalism is always a game of catch-up; every loan is made with the promise of additional sales income returning as a result. Across the economy as a whole, this promise has to be delivered for there to be a loan the next time, and the next time. In order to grow, the economy needs to go further into debt; meaning it must keep growing in order for increasing debt repayments to be made.

New research on capitalistic markets in a post-growth economy

Recently Tim Jackson and Peter Victor are among a number of heterodox economists who have begun to examine critically the last of the arguments above: *that banks' creation of interest-bearing credit in itself gives rise to a growth imperative.* Their position is that this argument is confused and its conclusion invalid. Taken together, these economists acknowledge that firms *are* reliant on credit to meet their costs of production in advance of sales; and that in extending loans banks are always created money *ex nihilo*. However, they state that, providing certain specific conditions were applied, it would be theoretically possible for such a system to exist in a non-growth state. Jackson and Victor (2015, 44) sum up the overall conclusion arising from such work: 'Contrary to claims in the literature, we found no evidence of a growth imperative arising from the existence of a debt-based money system per se.'

One argument here focuses on the distinction between the overall stock of debt in the economy, and the flows of repayments made in instalments. As the post-Keynesians Louison Cahen-Frouot and Marc Lavoie (2016, 165) write: 'It seems [...] that some ecological economists dealing with monetary issues somehow confuse stocks and flows [...] What has to remain constant is the stock, namely the debt, but the flow, namely the interest payment, does not need to be set to zero. Consequently debt-free money is not necessary [for a post-growth economy].' This is to say, it is theoretically possible to achieve a steady state economy with a stationary level of debt, even with banks continuing to issue loans to finance firms' production, so long as new issuances of credit were balanced out by repayments (and thus the destruction) of existing debt. The two further requirements in this case would be: first, that the flows of loan repayments were sufficiently rapid as to keep pace with the interest rate, thus avoiding the growth of compound interest and an increase in the stock of debt; and, second, that the interest payments received by the banks were then spent by the banks back into the economy again, either by being paid in wages to bank staff or in dividends to shareholders.

The fundamental basis of these arguments as to the compatibility of post-growth economics and interest-bearing debt is the post-Keynesian insight that the creation of credit, and with it interest-bearing money, responds to the demand from firms rather than the supply from banks. This is to say, it is the real economy which is in the driving seat. Extension of credit, while necessary for growth in economic production, is not the cause of it. Rather, it is the drive for growth, profits, and power which drives the demand for credit, and hence the increase in the overall stock of debt. If business's desire for growth could be constrained, the mere presence of banks lending them money at interest (e.g. to enable them to pay wages and suppliers in advance of sales, or to replace machinery whose costs would be paid off over the long term) would not in itself give rise to a growth imperative for the whole economy (Cahen-Fourot and Lavoie 2016, 165–167).

According to this understanding, a non-growth economy could retain *some* of the essential forms of a capitalist economy as we have long recognised it. On the one hand, Adam Barrett (2018, 235), another who has modelled the presence of interest-bearing debt in a post-growth economy, suggests that: 'Of course, in a zero-growth economy, there will still be some businesses that grow alongside others that shrink, and the dynamics of economic transformation and creative destruction will still occur'. On the other hand, zero-growth would also mean zero-sum, however, in the sense that, as Cahen-Frouot and Lavoie (2016, 167) put it, any growth in 'net saving and net investment by some households and firms could only occur at the expense of other households and firms.' For this reason, it has been suggested that it would become more difficult for start-ups to enter the market, and the pace of innovation could thus be expected to slow (Spangenberg 2010, 565). While there would still be room for the owners of capital to enjoy a stream of profits, this would have to remain within definite limits: as another analyst has suggested, 'those who want to continue earning income from surplus value [need] to somehow stop accumulation altogether and settle for a constant surplus' (Blauwhof 2012, 259). Or as Barrett (2018, 235) puts it, 'firm owners' income must be considered as either negligible or simply part of the wage bill.' Capitalists could remain, but in effect have to become wage-earners like other workers.

In this sense, while many of the forms of a capitalism economy *could* be retained within this model, it would still have to be a very different system from that which has been known under that title from its inception. Given these requirements, could such an economy still be described as capitalist? Even among those who have built such a model, some think not. Cahen-Frouot and Lavoie (2016, 167), for example, are clear: 'In our full stationary economy, [....] there is no accumulation of private wealth. Assuming private wealth is tantamount to capital in a broad sense, no further accumulation of capital by the private sector as a whole occurs. [...] Consequently, it would not be a capitalist system.'

Concluding discussion: capitalism without the capitalists

We return at the end to the division we began with. It does seem possible to imagine capitalism's survival through a transition to a post-growth economy—but only if one defines it in formal terms, as a set of features relating to private ownership and market distribution. If one understands the drive for endless accumulation to be the essence of this system, however, then the answer is clearly opposite. As the Marxist economist Minqi Li (2007, 29) has written, if a non-growth economy means not being able to accumulate capital, 'then what's the point of being a capitalist?' If capitalism could survive the transition to a post-growth economy, then, it would be a strange beast: capitalism without the capitalists.

In this light, the significance of the modelling which shows the compatibility of post-growth economics and capitalism could appear to be easily exaggerated. So many special conditions are required to enable this model to function as surely to reduce the significance of its contribution to this debate, its suggestion as to the survivability of capitalistic organisation in a post-growth economy appearing only trivially true.

Regarded another way, however, this theoretical work *can* be understood as a significant breakthrough; and one which, if it does not win the debate as to the sustainability of capitalism, points towards its transcendence. Let us take a moment to examine what this modelling does and what kind of economy it depicts. Its primary contribution is to offer a sophisticated model of a non-growing economy that is macroeconomically stable. This in itself is an important advance, since it suggests it is at least theoretically possible to transition to a post-growth economy without an economic (and thus social) breakdown. It offers, then, the vision of a viable economy which, in not growing, meets an essential precondition for humanity's having a hope of avoiding ecological disaster.

In addition to this, it suggests that this may be achieved while (indeed, depends on) retaining many familiar features of capitalistic organisation. This in itself can be viewed as a double strength. First, as Boonstra and Joose (2013) suggest, it is profoundly ahistorical to imagine the possibility of transition to an utterly different economic order. Rather than being inscribed on a blank slate, they argue that it is only possible for a new order to emerge from a preceding order, in which it has been developing in embryo. In this sense, they suggest that, in whatever form a post-growth economy would emerge, it and its capitalist predecessor ought to be viewed as diachronically separated developments of the same state.

A second strength would inhere, not in the sheer fact of continuity with capitalistic forms but with the qualities of those forms in themselves. This is to say, the post-growth model suggested as possible by the likes of Jackson and Victor features the freedom and flexibility over production that comes with the existence of a plurality of independent employers (which could be a mixture of privately-owned firms, co-operatives, and some form of shared ownership hybrids). It is a wage economy, meaning it allows for the diversity and productivity of job specialisation, and can provide flexibility for both workers and employers as to where to work and whom to hire or work for. Meanwhile, as a monetary economy it uses markets and price signals to connect consumer preferences with production standards and options. In its best sense, this form of economic organisation manifests virtues of both efficiency and freedom. In a context in which this model is advanced as a viable alternative principally to a business-as-usual capitalism which promises ecological disaster, but also stands as an alternative to political visions which seek to arrest ecological dev- astation but only at the costs of authoritarianism, its virtues ought to shine even brighter.[3]

Perhaps the notion of 'capitalism without the capitalists', far from being a rather embarrassing contradiction in terms, represents a theoretical strength. One thing it resembles, in fact, is a conception of 'markets without capitalism' which characterised the idea of market socialism as developed in the UK in the 1980s and 1990s by Julian Le Grand and others (Le Grand and Estrin 1989). These ideas developed in response to both the failures of state socialism and the neoliberal critiques of Keynesian statism. Here, too, was a conception of a new type of economy in which the forms of capitalistic organisation were retained and put to work, via a framework established by the state, towards the achievement of ethically-selected ends. This vision was depicted as having a number of cardinal virtues—equality, liberty, efficiency, and practicality— which promised advantages over both capitalism and state socialism. Against capitalism, it promised not just greater equality (via the state's intervention to ensure that disparities in opportunities and outcomes were constrained within set limits), but greater freedom, in the sense of the positive liberty that derived from a greater sharing of the resources required for self-development. Against state socialism it also promised greater liberty, in the sense not only of freedoms and flexibilities within the economic sphere, but also in the sense of promoting an independence and diversity of political views (compared to a situation in which the state was the sole employer or direct bankroller of every organisation). Alongside this were promises of efficiency, through allowing the market to co-ordinate production decisions in a decentralised manner, and practicality, in the sense of relying on established means of provisioning large, complex societies. The latter virtue was contrasted in particular with the radical alternative to abolishing markets within a communist endstate, about which Marx was always vague, thus leaving no actual economic theory of a communist society.

Perhaps now the confusion of Daly's remarks about the political econ- omy of a post-growth society—against both capitalism and socialism, but for a system of markets and private ownership that could also be described as socialist—begins to make sense. Perhaps what we can see in this model is the beginnings of a convergence of capitalism and socialism, within the terms

---

3 For a discussion of some authoritarian scenarios of political reckonings with climate change see Mann and Wainright (2018).

set by environmentalism.[4] Perhaps, that is, this represents the collapsing of economic alternatives into a synthesised vision under the mutual pressures of our environmental and social realities. Once you combine our need to rapidly constrain our exploitation of the environment to bend within key planetary boundaries, with our need insofar as we can to go on living in a certain fashion—as highly populous, diverse, technologically advanced societies—then the possibilities for economic organisation take on a certain shape. Abiding by environmental limits, abolishing the systemic drive for accumulation, respecting the internal dynamics of macroeconomic stability, and preserving a sense of pluralism, freedom, and dynamism—these are the defining conditions for any viable economy from this point forwards. On a theoretical plane, perhaps we could also see here the prospect of a knitting together of various strands of heterodox economics—ecological economics, Marxist political economy, and post-Keynesian economics—into a consistent stream. Politically, perhaps this means Kunkel's reference to a 'steady-state social democracy' best describes the spirit, mixing ideology and pragmatism, suited to forming an effective political movement around this economics.

We began with two questions, not only 'Does transition to a post-growth economy mean abandoning capitalism?' but also 'Does this matter?' What answers could we give in respect of the post-growth model discussed here, in which capitalism and socialism could be said to achieve some form of convergence?

Is capitalism compatible with this vision?—Sort of. Does this matter?—In the way that it is compatible, very much so.

Regarding the political reception of this concept of post-growth economics, there are clear potential advantages to be had. On a tokenistic level this blurring of its political lines might disrupt the automatically negative reactions that a simple identification either with capitalism or socialism might variously draw from left or right. More substantially, its reception might itself be aided by the theoretical design of a post-growth economy. That is to say, this design combines virtues of freedom and social protection, and does so while establishing lines of continuity with our economic reality as we already understand it. There is a hint of pragmatism here, which in itself may give some grounds for hope.

---

4 Daly's remarks about a post-growth economy as 'a Jeffersonian-type, small-scale capitalism' which might also resemble conceptions of 'eco-socialism' find some echoes in some contemporary left-wing visions of high-tech but decentralised economies, comprising small firms and co-operatives. See, for example, Graeber (2016) and Mason (2015).

Acknowledgements

I would like to thank John Foster, Simon Mair, Andrew Jackson and Tim Jackson for their comments on a draft of this chapter. Any disputable points of argument and interpretation remain my own work.

Chapter 2

FACING UP TO CLIMATE REALITY: INTERNATIONAL RELATIONS AS (UN)USUAL

Peter Newell

Introduction

Imagining a world political order at +4°C in any detail is not an easy, enviable or even viable task, given the lack of historical precedents to draw on and the multiple uncertainties and unknowns we face. These include economic and ecological uncertainties about the future, and about the pace of political change in the contemporary world order introduced by rising nationalism and populism in recent years with their rejection of supranational 'liberal' regional and global governance.

Yet this does not do away with the need to consider and anticipate potential scenarios, and to pre-empt and manage the sorts of challenges the international system is likely to face in a world of accelerated or even runaway global warming. Some of these challenges can be read off or scaled up from current models and forecasts of expected climate impacts from the IPCC and other bodies, as well as the experience today of the effects of climate change on politics at all levels (including the global) and upon the lives of vulnerable groups in particular, which are set to magnify and intensify substantially. Others can be extrapolated from how the international system, and more importantly and pertinently, the most powerful actors within it, have reacted to previous crises (albeit often not environmental in origin) combined with an assessment of their willingness and ability to do so in today's world given where the political centres of gravity lie.

It is important to note at the outset that though this book takes as given the reality and severity of climate change, and can justify such concern with reference to a near unanimous scientific consensus about the scale of the threat posed by anthropogenic climate change, most of the world's population appears not to. That indeed is the problem. The question then is *whose* and *which* climate reality are we talking about? We inhabit a world of everyday climate denial which proceeds as if climate change is not an imminent threat. Everyday climate denial is omnipresent in decisions to expand airports, extract new fossil fuels and pursue business-as-usual growth, including the financing and promotion of fossil fuel intensive development models in the global South by powerful international institutions such as the World Bank (Oil Change International 2016). There are many reasons for this denial, including our collective psychological inability or disinclination to process the scale of the threat and its dystopian implications (Hamilton 2010), the failure of media with ties through advertising to fossil fuel industries and large consumers of fossil fuels (such as car companies and airlines) to communicate the extent of the threat (Newell 2000), and the power of incumbent interests in business and government who profit from unsustainable development to ignore, negate and play down the implications of the threat it poses to societal well-being and our planetary survival (Newell and Paterson 1998).

Recognising this situation is not to overlook the ways in which some ruling elites and establishment actors have engaged with the severity and 'reality' of climate change in ways I explore below. This includes not only international institutions, such as those charged with responding to climate change, but also transnational business actors (Newell and Paterson 2010) as well as military actors (Buxton and Hayes 2016). But they have largely done so with a view to securing for themselves new institutional mandates, opportunities for profit from the climate crisis and new channels of resources for militarism, rather than to engage in more progressive and transformative forms of politics. Having looked at these processes of accommodation of 'climate reality' by actors who face a legitimacy crisis precisely because of their role in driving climate change, I then look at the ways in which climate change may be disruptive of the 'normal' conduct of international relations (IR). I explore how IR cannot be the same in a warming world. Climate change will draw attention to itself both through spectacular disasters such as droughts, floods and extreme weather events, as well as through the slower-moving but equally lethal disruption and destruction of the water, land and atmospheric systems which sustain life on earth (Nixon 2012). This will require major international responses involving unprecedented cooperation and resource demands, as well as potentially heightened conflict and insecurity induced by tensions over affected resources or as a result of population displacement in politically volatile regions. An almost permanent crisis mode can be expected to become the norm for the institutions of global governance. But as governments face increasing demands on their resources and tests of their ability to protect their citizens, it is far from guaranteed that responses will be guided by a new ethic of humanitarianism as opposed to a lifeboat ethic (Hardin 1968). We know from bitter experience that disasters and catastrophes are as likely to be used as opportunities to advance and entrench socially regressive forms of politics and unsustainable trajectories as the reverse, as Naomi Klein's (2007) book *The Shock Doctrine* shows so clearly.

Growth will become increasingly difficult, if not impossible, to secure because of the costs of damage to infrastructures, human and ecosystem health, impacts on key sectors such as energy and agriculture and demands for funds for relief and compensation, as we are already observing in discussions around loss and damage in the climate change negotiations. I will also explore how the development sector will have to reorient itself, the likely shifts in global governance required to deal with this, and what this means for renewed militarism. I then provide a more positive reading of the ways in which the new climate reality and the crisis of industrialism which it intensifies could lead to more progressive moves towards a sustainable society. I conclude with some thoughts on how to engage with the severity of the situation without deepening disempowerment, disengagement and regressive politics through an authoritarian and anti-democratic politics of top-down control from the very elites that are largely responsible for the runaway climate change we now face.

Crisis? What crisis? International Relations as usual

We need to be aware of the chasm between the climate reality that Greens take as given—one which reads from the science the need to radically overhaul the economy, even potentially the capitalist system in its entirety, and to pursue a project of de-growth (Blewitt and Cunningham 2014; D'Alisa et al 2014)—and that, on the other hand, of the elites which, when they consider climate change at all, gives rise to attempts to manage climate change mitigation and adaptation in ways which are consistent with, aligned with, and subservient to the need to sustain and expand growth at all costs. Incumbent power has been employed not only to promote the possibility of 'green growth' (OECD 2011; World Bank 2012) and 'climate compatible development' (Nunnan 2017), but also to maintain and even extend the reach of militarism in society (Buxton and Hayes 2016). Thus, events, disasters, trends and projections which might from a Green point of view imply the obvious need to 'face up to climate reality' and to fundamentally transform the ways in which global politics is conducted, end up being pressed into the service of the business-as-usual pursuit of industrial growth on behalf of the military-industrial complex (Koistinen 1980).

Taking 'climate reality' as a starting point does not automatically lead to positive or progressive action. Consider the responses to the Paris Agreement in the relatively progressive space of the UN climate negotiations, where one might imagine the reality of climate change is taken most seriously. Many of the ambitions of the 2015 UNFCCC Paris agreement about net-zero emissions, for example, imply the widespread use of NETs (negative emission technologies), the most commonly proposed form of which are BECCS (Biomass Energy with Carbon Capture and Storage), utilised in more than 80% of IPCC pathway projections. BECCS involves the mass planting of trees to absorb carbon dioxide from the atmosphere. Even disregarding the technological issues involved here, for these to work at the scale required means that plantations three times the size of India, consuming one third of the planet's arable land, would need to be created (Anderson and Peters 2016). Silences around the viability of such assumptions (given not least the competition over land and resulting social dislocation they would presage) are deafening. The reality of failed climate mitigation is the starting point for the promotion of CCS, carbon dioxide removal technologies and geoengineering, and reflects the reluctance of powerful states and corporations to contemplate the kind of economic restructuring required adequately to address climate change. Without challenging and disrupting power relations, appeals to act on the urgency of climate change can lead to these sorts of regressive responses.

The securitisation of climate change

The threat climate change poses to national and international security offers a compelling case of how powerful narratives, however well-grounded in science, are often deployed in politically problematic, counter-productive and regressive ways. Let's take the case of climate wars. From invoking fears about waves of climate refugees moving from new arenas of conflict or areas of the world increasingly uninhabitable because of climate change, to projecting 'climate wars' over diminishing supplies of water or over remaining oil reserves, some in the environmental movement have been willing to promote simplistic neo-Malthusian narratives about scarcity-induced conflict. The blurb on the cover of Harald Welzer's (2012) book *Climate Wars* reads, for example: 'Struggles over drinking water, new outbreaks of mass violence, ethnic cleansing, civil wars in the earth's poorest countries, endless flows of refugees: these are the new conflicts and forces shaping the world of the twenty-first century'.

Indeed, some scholarly research has documented environment-induced conflict and sought to isolate primary drivers (Homer-Dixon 1999),[1] while NGOs such as International Alert have compiled lists of states they claim are at risk of armed conflict because of climate change, including Somalia, Nigeria, Colombia, Indonesia, Algeria and Iran.

Pronouncements from leading figures in international relations illustrate the growing recognition that environmental degradation poses a threat to national and international security. UN Secretary General Ban Ki-Moon addressed the UN General Assembly about water wars in reference to crises in Kenya, Chad and Darfur, labelled the world's 'first climate change war'. The United Nations Security Council (UNSC) has held two debates on the international security implications of climate change in recent years (2007 and 2011), the UN General Assembly (UNGA 2009) commissioned a report on this issue in 2009, while the security threats associated with global climate change have also been identified and explored by the UN Environment Programme (UNEP 2007), the UN Development Program (UNDP 2007), and the UN Secretary General (Moon 2007). Regional organisations from the European Union to the Pacific Islands Forum have identified climate change as a current and growing security threat, while climate change has found its way into national security statements of key political institutions throughout the world, from the USA to the UK, Australia, Russia, Finland and Germany, among many others (McDonald 2013).

Yet while the idea of climate change as a security threat is gaining both academic and practical purchase, important differences in the logic of this link suggest radically different responses to climate change as a security concern. The statements above from leading organisations come on the back of growing interest from the military establishment in the threat posed by environmental factors to national and international security through a variety of direct and indirect channels, such as flows of refugees from flooded areas, intensification of conflicts over oil and water, health security pandemics caused by environmental factors (malaria, spread of disease), and disruptions to infrastructure (stresses upon transport systems, disruption to food supplies and the ensuing potential for civil unrest). In turn, this has given rise to a series of critiques of resource determinism and neo-Malthusian narratives around conflicts over diminishing resources in 'climate war' scenarios, as well as warnings against securitising environmental threats by constructing them in this way. Research positing direct and linear causal connections between climate and war has

1 For example Marc Levy et al suggest in relation to intrastate conflict that 'severe, prolonged droughts are the strongest indicator of high-intensity conflict' (conflicts involving more than 1,000 battle deaths). CIESIN www.ciesin.org/levy.html.

been critiqued on a series of grounds, including: that the correlations identified are spurious since they always rest upon causal assumptions which range from the arbitrary to the untenable; that even if the correlations identified were significant and meaningful, they would still not constitute a sound basis for making predictions about the conflict impacts of climate change; and that such models reflect and reproduce a problematic ensemble of Northern stereotypes, ideologies and policy agendas (Selby 2014).

Such narratives are potentially attractive to military actors keen to find new roles for themselves, utilising the financial resources this would give them access to as protectors of supplies of water and land around the world: the so-called 'green beret' phenomenon (Elliott 2004; Eckersley 2004). This is the danger of framing climate change even as a 'threat multiplier' (rather than mono-causal driver as in some early accounts): that it implies and validates a need for military responses to contain threats to security that bypass public political discussion and potential contestation in favour of 'high politics' (Newell and Lane 2018). Positing environmental threats such as climate change as security threats could encourage a military response: a response inconsistent with both the requirements for an effective solution to problems of environmental change and with proponents' goals of challenging existing discourses of security in global politics (McDonald 2013).

It is precisely this kind of interest on the part of the military establishment of which Greens are generally both suspicious and critical, and which sparks concern about linking these issues. In their book *The Secure and the Dispossessed*, Buxton and Hayes (2016) show 'how the military and corporations plan to maintain control in a world reshaped by climate change. With one eye on the scientific evidence and the other on their global assets, dystopian preparations by the powerful are already fuelling militarised responses to the unfolding climate crises'. In a more measured fashion, McDonald also notes that 'the most powerful discourses of climate security are unlikely to inform a progressive or effective response to global climate change' (2013:42). Militarism is often mobilised to protect the 'secure' and their assets and not the dispossessed or even the losers from climate change who in their role as migrants and refugees are constructed as threats (Buxton and Hayes 2016). McDonald reveals how a 2003 Pentagon report proposed that some states 'might seek to develop more effective border control strategies to ensure that large populations displaced by manifestations of climate change (whether rising sea levels or extreme weather events) could be kept on the other side of the national border', such that 'people displaced by environmental disasters or environmental stress may be positioned as threats to the security of the state rather than as those in need of being secured' (2013:46). Dominant framings are produced by actors with a stake in protecting or expanding expenditure in their sector, and who benefit from threat proliferation, which justifies their existence and indeed growth. The pitch to policy-makers is around climate adaptation and their ability to secure assets and infrastructures and protect borders. Climate mitigation is understandably ignored, given the vast contribution of the military to global GHG emissions.[2]

Closely related to securitisation narratives are claims about (climate) refugees which invoke powerful imagery to stoke nationalist fears about 'swarms' of immigrants from conflict zones or areas no longer habitable in a warming world, often overlooking the role of industrialised countries in causing that population displacement through climate change and war. As with security, while at times these narratives are invoked to stimulate action, the consequences, when left to powerful incumbent actors, can often be regressive. The development sector would be centre stage in such a scenario, called upon by global governance institutions to work in this 'humanitarian' space in the wake of these 'natural' disasters. In a more dystopian vision, there could be scope for renewed militarism in moments of crisis and emergency and demands for ever-increasing 'states of exception' which bypass politics as normal.

Climate compatible growth and development

If security is traditionally considered to be the primary and overriding concern of states in an anarchical society characterised by the absence of a central overarching authority (Bull 1977), the economy and the pursuit of growth surely follows close behind. The indictment of the project of infinite growth represented by runaway climate change poses another challenge for global elites to manage. As Newell and Lane (2018) suggest:

> This suggests the need to understand how the geophysical, social and political dimensions of climate change interact with incumbent political and economic systems... At times climate change disrupts them and creates crises for the organisation and legitimacy of existing ways of ordering the world, revealing systemic vulnerabilities (e.g. the dependence on finite resources for the current production of food and energy or for trade and transportation). At other times opportunities are projected onto it to 'creatively destruct' in Schumpeterian terms: to make money from crisis or to organise new rounds of accumulation around 'climate compatible growth'.

2 For example, the U.S. military alone uses more oil than any other institution in the world. Every year US armed forces consume more than 100 million barrels of oil to power ships, vehicles, aircraft, and ground operations (Union of Concerned Scientists 2017).

As well as representing a threat to the current fossil fuel-powered global geopolitical configuration, against which fossil fuel interests have sought vigorously to defend themselves (Newell and Paterson 1998), climate change has also been repositioned as an opportunity. This includes attempts to reconcile climate change with capitalism through narratives around green growth (Bailey et al 2011, Wanner 2015), as well as proposals for a fundamental diversion of technology, finance and production towards low carbon goals (Newell and Paterson 2010). From the OECD to the World Bank, prominent global governance institutions have been at pains to show that continued economic growth, and the energy required to power this, is not only compatible with tackling climate change, but is a prerequisite to tackling the issue given the requirements for finance, technology and new forms of production (OECD 2011; World Bank 2012). This focus on expansion of supply, rather than reduction in demand, upon changing patterns of production and consumption but not levels, even among governments that accept the need to live within 'planetary boundaries', goes to the heart of some of the contradictions facing states in a growth-oriented capitalist global political economy. These arguments rely on an absolute decoupling of economic growth from energy and material throughput which, while frequently asserted (e.g. Handrich et al 2015; UNEP 2015), is now widely considered to be impossible (Giljum et al 2014; Knight and Schor 2014; von Weizsäcker et al 2014; Mir and Storms 2016; Ward et al. 2016; Schandl et al 2017). The analysis from the *Working Group on Climate Change and Development* suggests that growth in OECD countries cannot be squared with halting warming at 2°C, 3°C or even 4°C (NEF 2009).

In a related way, the development industry has sought to respond to climate change by renewing its claim to be a saviour of the poor (Newell 2004; World Bank 2010) while ignoring its own complicity in driving climate change, both historically and to this day, through its ongoing promotion of fossil-fuelled development (Oil Change International 2016). The World Bank, for example, continues to provide high levels of finance to fossil fuels, indeed doubling its funding for fossil fuels between 2011 and 2015. It provided USD 1.7 billion in total investments for exploration or projects that included an exploration component during these years. Climate-compatible development has emerged as the preferred narrative for articulating the claimed 'triple wins' possible from projects and interventions that simultaneously reduce emissions, enhance capacity to adapt to the effects of climate change, and help to alleviate poverty through growth. One expression of this is 'climate smart agriculture', which seeks to displace the focus from the effects of industrial agriculture on the climate system and onto the potential for 'sustainable intensification' of output through GM crops, biochar, biofuels and the like, while obscuring the effects of fertiliser production and use (Newell and Taylor 2018). As critical writers in development have long noted, export-led, private-led capitalist industrial development is the answer to whichever way you pose the development question. As Crush (1995:10) notes, 'development is always the cure, never the cause'. It is a given of the development industry, and questioning of it is off limits. As a World Bank official told me privately: despite increasing interest in ideas about post-growth, de-growth or prosperity without growth (Jackson 2011), there is no space seriously to take on the significance of these issues in a neoliberal institution whose funding comes from the world's richest capitalist countries.

International Relations, but not as we know it

Despite the evidence presented above of successful attempts at accommodation, denial and dilution of the threat posed by climate change to the everyday conduct of IR around security, economy and development, some things will not remain the same and will generate more disruptive change.

Post-growth economies might emerge by default rather than design, as a result of the rising costs of protecting and repairing infrastructure (from flooding, heat stresses and other weather related damage); rising health costs (heat waves, disease, malnutrition, death from 'natural' disasters); social costs (compensating workers/sectors hit by mitigating and adapting to climate change and impacts on productivity in key sectors such as agriculture, particularly in the global South); propping up failing energy systems as their utility and efficiency declines, shortfalls appear and they collapse under pressures of demand or pressures such as heat stress. In sum, state expenditures and costs to private actors will, at some point, outstrip their ability to generate new growth.

Given the largely unquestioned status of economic growth as a societal goal and benchmark of political success, the inability of states to deliver on this goal over time could generate considerable unrest and disruption, as well as legitimacy crises, intensifying the threats democracies are increasingly facing the world over. Economic historian Benjamin Friedman, in his (2006) book *The Moral Consequences of Economic Growth*, identifies a clear pattern in the historical relationship between economic growth and social values. During periods of rising economic prosperity, people tend to be more tolerant, optimistic, and egalitarian. Periods of stagnation and recession, by contrast, have been characterised by pessimism, nostalgia, xenophobia, and violence. During times of scarcity, people are more likely to look for scapegoats than to pull together, more prone to zero-sum thinking, and more susceptible to the appeals of populists and demagogues.

Added to this is the fact that the levels of growth experienced to date probably constitute an historical anomaly and are unlikely to be repeatable. For Dobson (2014:8), 'Growth, the unexamined assumption that underpins our current political settlement, is nearing its sell-by date'. He suggests 'the past 250 years have been an era of exception, rather than normality and we believe that this era of exception is coming to an end, with potentially calamitous consequences, but also potentially liberating implications'; and that 'we have come to think of the industrial era of Promethean expectation and performance as normality, whereas it is in fact a world-historical era of exception' (2014: 25). For some, this will occur as limits are placed upon the use of fossil fuels or because of the impacts of climate change (causing negative growth), such that more and more state and private investment will be required to prop up existing infrastructure, to pay for emergencies and losses through health and 'natural' disasters as noted above. We will, in other words, be running to stand still economically. Forward projections of growth spell out the challenge of sustaining it. When politicians aim to deliver annual growth rates of over 2% per year (much more in many contexts), this implies an expansion of the economy by a factor of 10 every 100 years; or that in 200 years the economy will be 100 times bigger than it is now.

The challenge then becomes that of how to move towards managed transition rather than reactively governing emergencies (Dobson 2014) of the sort we see with increasing frequency—whether civil unrest around water shortages in Cape Town, flooding in New Orleans or Mozambique or food riots in Mexico prompted by the rising price of corn, triggered in turn by the biofuel boom (Smith 2010). The challenge of managing a transition applies within countries as well as around the global displacement of risks and costs. We can see already the tendency to exploit global inequalities and patterns of uneven development by employing a range of spatial and temporal fixes to displace and move problems around (Harvey 1981), given capitalism's inability to solve the underlying contradictions that they represent (O'Connor 1994). This includes the promotion of carbon trading, which enables richer polluters to buy permits generated from projects in poorer parts of the world rather than reduce their own emissions at source (Lohmann 2006), pushing the costs of adjustment onto others and into the future, what Bumpus and Liverman call 'accumulation by decarbonisation' (2008). The promotion of biofuels in the global South to provide energy for transport in the global North and the waves of 'green grabs' (Fairhead et al 2012) to acquire land to meet the future developmental and environmental needs of the rich core of the global economy are other examples of 'spatial' and 'temporal' fixes in operation.

Under this scenario, the international system and the institutions of global governance within it would be faced not only with waves of humanitarian crisis and social and economic breakdown, but also with the need for global oversight over more risky strategies around geoengineering and techno-fixes in the face of continued failure to bring down emissions and of the intensification of conflicts resulting from displacement and dispossession of communities in the rush to 'grab' land and other resources. Powerful states will be under increasing pressure to assume moral leadership to protect those most affected by climate change—but may also, given the other pressures described here, adopt Hardin's famous 'lifeboat ethics' (1968). 'Disaster collectivism' in communities may become the default for many communities in the absence of state intervention on their behalf.

One scenario might be that system vulnerabilities and resource dependencies prompt a move towards greater self-reliance and self-sufficiency (just as interest and investment in energy efficiency and renewable energy increased in the wake of the OPEC crisis). While the end of oil may shift the terrain of conflict, since renewable energy systems are more locally owned, we could also of course see an intensification of conflict for some rare earth minerals, and materials such as lithium, required for electric car batteries (Newell and Mulvaney 2012). The growing cost of imported food (because of rising transport costs, crop and pest damage induced shortages and scarcity) could drive a re-emphasis on local and bio-regional food production. But reduced availability of food (for the same reasons) might also exacerbate conflict and social tensions. Again, more positively, resource constraints (whether imposed from above, through the market, or by the effects of climate change itself) might create momentum for a 'conserver society' (Trainer 1996) of sharing, repairing, re-use and recycling. This may trigger or require a reconfiguring of work; a shorter working week and job-sharing to reduce impact and share the available work.

State action will clearly be a critical part of societal responses to climate crises. Whilst suggestions and speculation about the ways in which the climate crisis could provide an opening for more progressive responses might understandably prompt scepticism about the likelihood, willingness and ability of states in a neoliberal global economy intervening in the sorts of ways called for to tackle climate change, it is worth recalling that state action has brought about radical, progressive, rapid and disruptive change in the past and could do so again today (Simms 2013; Simms and Newell 2017). Think of the New Deal in post-World War II America, strategies of wholesale industrial

45

conversion in the UK, or state-led 'energy revolutions' in Cuba (Simms and Newell 2017). More recently and encouragingly, there have been recent bold moves by governments to leave fossil fuels in the ground. A combination of divestment of finance from fossil fuels by major investors and laws and regulations that many governments have recently shown themselves willing to adopt to keep fossil fuels in the ground (such as recent moratoria on new oil exploration and production announced in 2017 and 2018 by a number of countries including New Zealand, France, Costa Rica and Belize), or which set clear near-term timetables for their phase out as is happening in China, shows what is possible. Costa Rica, for example, has a moratorium on oil exploration in place that was recently extended to 2021, the year by which Costa Rica intends to be carbon neutral as well. There are also moratoria on fracking in a number of jurisdictions globally such as France, Germany, Ireland, Wales, Scotland and Uruguay.

Ideologies and ideas of what states can and should do, the degree of policy autonomy and developmental space they have been able to enjoy, are not fixed (Johnstone and Newell 2017). They are negotiated and contingent on the weight of historical forces at key moments. Hence, after decades of criticism of the efficiency and effectiveness of the state, the prevailing orthodoxy today is that states can and must play a proactive part in leading transitions to sustainability.

International Relations, as we may want it

Having discussed how growing recognition of the severity of climate change is leading key actors in IR to accommodate, dilute and dismiss calls for more radical transformation in the face of climate change (which we can observe in the processes of continuity as well as change and disruption described above and foreseeable in the future), here I look at opportunities to re-make and re-order IR in more progressive and ecological ways. It raises the question of whether it is possible that 'shock doctrines' (Klein 2007) and moments of crisis could create openings to move towards a more sustainable society.

Firstly, the onset of climate crisis may amplify calls for the key shifts required to bring the mandates and powers of institutions in line with the new climate reality. For example, the World Trade Organisation (WTO) could take on a stronger mandate to pursue sustainable development, allowing exemptions from normal trading rules for policies and measures aimed at tackling climate change (consistent in aim at least with World Bank calls for a new 'energy round') or a range of other environmental threats. Or it could be redesigned, along the lines proposed by Molly Scott Cato and others, as a body for International Trade and Sustainable Development (LeQuesne 1995; Cato u.d.). Fundamental reform of the IMF and World Bank would also be required, if not the disbandment of those institutions, given that the narrow ideologies and policy proscriptions they advocate are incompatible with a more sustainable world. If suitable reforms are achieved, these institutions, which have invested so heavily in fossil fuel infrastructures to date, could play a central role in financing the transition through taxes on fossil fuels and the redistribution of fossil fuel subsidies alongside the Green Climate Fund and Adaptation Fund.

In the climate change negotiations, the need to forge a new global settlement could result in adoption of the principle of 'contraction and convergence', as promoted by the Global Commons Institute and supported by a wide range of developing countries in particular. Placing equity centrally might also mean adopting the Greenhouse Development Rights framework (GDR 2018), which seeks to tie obligations to reduce emissions to levels of development across society rather than basing them on nation states, which obscures who consumes and produces most greenhouse gas emissions. There might also be potential for alliances within countries and globally between beneficiaries from action on climate change and the many victims of climate change impacts. A key driver here would be that, as countries fail to deliver on mitigation, the costs of adaptation rise significantly (Stern 2008). These pressures would be amplified by the growing chorus of demands for compensation for 'loss and damage' and waves of human rights claims and climate change litigation (Humphrey 2009).

This could be accompanied by non-proliferation treaties for fossil fuels, organised along the lines of existing nuclear non-proliferation treaties (Simms and Newell 2018), with proper mechanisms for mutual monitoring and reporting. A clearer role might be envisaged for the UN Security Council in condemning and sanctioning state actions that violate climate norms, boundaries and targets. Alongside this, there is a need for clear principles of transnational harm and liability for environmental negligence beyond state borders (Mason 2005; Eckersley 2004). This would include strengthened corporate liability regimes, including Foreign Direct Liability in order to hold companies to account for common and universal responsibilities when operating abroad (Ward 2002; Newell 2001). More fundamentally, it would require an overhaul of the priorities that currently govern global politics. Susan George in her book *Whose Crisis, Whose Future?* proposes reversing the current 'spheres of priorities' in today's financialised world of global politics in the following way:

47

Our beautiful finite planet and its biosphere ought to be the outer-most sphere because the state of the earth ultimately encompasses and determines the state of all other spheres within. Next should be human society, which must respect the laws and the limits of the biosphere but should otherwise be free to choose democratically the social organi-zation that best suits the needs of its members. The third sphere, the economy, would figure merely as one aspect of social life, providing for the production and distribution of the concrete means of society's existence; it should be subservient to, and chosen by, society so as to serve its needs. Finally, and least important would come the fourth and innermost sphere of finance, only one among many tools at the service of the economy (2010:3).

This has some resonance of course with Kate Raworth's (2017) ideas about 'doughnut economics', meaning that a safe social operating space for human-ity needs to be accommodated within planetary boundaries. The concept of the 'doughnut' describes a 'hole' of critical human deprivation in the middle below a social foundation representing the minimum amount of well-being for humanity. An ecological ceiling is an outer ring representing planetary resource limits, and between both of these a safe and just space for humanity. This suggests perhaps some basic design principles for a green global govern-ance able to deal with the reality of climate change, among other planetary threats.

To deliver on such an international system, procedural changes would also be required to democratise (as well as 'ecologise') global governance. Some Greens and academics have sought to identify and promote greater direct and indirect representation of citizens in global decision-making through citizen assemblies and the like (Stevenson and Dryzek 2014). In terms of the ques-tions of voice and representation, Greens have called for an ombudsperson for future generations (Read 2012). As part of their call for 'Planet Politics' Burke et al (2016: 516) meanwhile propose that 'it is time to consider whether major ecosystems—such as the Amazon basin, the Arctic and Antarctic, and the Pacific Ocean—should be given the status of nations in the UN General Assembly and other bodies, or new organisations established with the sole purpose of preserving their ecological integrity.' They suggest that 'voting rules and attitudes must change. We suggest the creation of an 'Earth System Council' with the task of action and warning—much like the current UN Security Council—that would operate on the basis of majority voting with representation of Earth system scientists, major ecosystems, species groups, and states' (Burke 2016: 516).

Consistent with arguments made throughout this chapter about the need to ensure that calls for political programmes based on 'climate reality' do not inadvertently entrench and reproduce regressive and anti-democratic politics, it is worth noting that Greens are not united in their view about the desirability of proposals, such as those described above, to strengthen 'governance from above'. Chandler et al (2017: 7) criticise proposals for a new 'Planet Politics' on the grounds that 'the implicit assumption that technocrats and advocacy groups can mobilise with only the need for a minority of states' support appears to provide a new legitimacy to global liberal "coalitions of the willing"'. Instituting global governance in 'firm and enforceable' ways, as if there were universal solutions that could be imposed from above, they suggest, 'is a recipe for authoritarianism and new hierarchies and exclusions'. In particular, they highlight a problematic 'desire to jump straight into the "limits to human freedom" on the basis that this is what "the planet" is telling us' (ibid: 8) without clarifying how 'the planet' coveys that message and who is 'us'.

Similar criticisms are made of calls for strengthened earth system governance (Biermann 2014) where governance is led by expert international institutions through the conclusion of environmental treaties. The claim is that political and ideological contestation over definitions, problem identification, desired outcomes and proposed mechanisms to achieve these are replaced by techno-managerial planning and decision making (Swyngedouw 2013), or what Hajer et al (2015) refer to as 'cockpit-ism': the illusion that top-down steering by governments and intergovernmental organisations alone can address global problems. The fear is that this clears the path for geoengineering and other forms of 'planetary management' (Stirling 2015). The challenge for Greens in many ways is to make use of the modelling of planetary boundaries and attempts to anticipate and avoid key tipping points in the earth system, building on the core ideas (as well as limitations) of limits to growth approaches, without indulging in a politics of global control and managerialism.

Conclusion

In many ways, the threats and challenges that climate change poses to the international system are a) without precedent in terms of their scale and scope for disruption; and b) impossible to gauge with any degree of precision, despite the best efforts of the scientific community. Nevertheless, both with reference to historical examples of how states and the international system have sought to manage and respond to crises, as well as to the nature of responses so far to the impact of climate change on key domains of global politics like security, economy and development, this chapter has explored the politics and prospects of IR today and of IR as it could be, if re-made and re-ordered in line with emerging climate realities.

There is a genuine question as to whether the liberal order can hold up under the strains of conflict, migration, economic dislocation and rising costs associated with the damage wrought by climate change. The task of building an international system able to address these realities is all the more daunting when one bears in mind that major reforms of global governance have in the past occurred in the face of visible and spectacular crises such as war and financial breakdown, and ones which affected predominantly the richer parts of the world, and less so in response either to crises which confronted poorer and marginalised groups with weak political representation, or which are characterised by 'slow violence' (Nixon 2011). The history of the Marshall Plan (the precursor of the OECD) shows that self-interest on the part of powerful states, usually also seeking to respond to the needs of powerful domestic capital, is often the key driver (Burnham 1990). Enlightened self-interest can extend to efforts such as those of the US to construct a new international economic order (in the form of the Bretton Woods system), to maintain an ostensibly liberal open trading system, and to provide assistance to countries facing short-term balance of payments crises (the rationale for creating the IMF). Pegging currencies to the dollar and seeking to keep trading channels open for the export of goods from a US economy less damaged by the experience of the Second World War provided plentiful incentive for the assumption of a leadership role.

The question then is, who and where are today's leaders driving changes in the global political order to align it with the goal of tackling climate change? Those states most affected by climate change—Pacific and low-lying island states, and some of the least developed countries in the world—have been vocal in adopting the language of human rights and even of crimes against humanity, but they remain weak and politically marginalised. Some European and other leaders have provided fluctuating leadership, but as the latest IPCC Special Report on 1.5°C shows, none has called for, let alone adopted, the sorts of changes or transformations compatible with keeping emissions below 1.5 degrees, nor with adapting adequately to the degree of climate change we are already committed to. In any case, state action, though decisive, will not be sufficient on its own, as recognised by the Paris Agreement's efforts to include the contributions of businesses, cities and civil society organisations. I have also suggested that in imagining alternative IRs and forms of global governance, we need to move beyond illusions of control from above. As Burke et al note, 'the gravity of the changes to the biosphere that climate change will wreak grant the climate an independent agency that will exceed the agency of any state, group, or the state system itself' (Burke et al 2016: 513). Responses from below, such as coping and self-help strategies, autonomous forms of mutual aid, and action to prepare for climate reality in the way that movements like Transition Towns are trying to do, are also sites of the global governance of climate change.

Given levels of denial to date, what will it take to make countries (and their citizens) face up to climate reality? Many assume that once richer countries experience the effects of climate change, action will follow. Yet the experience of hurricane Katrina in the USA, floods in the UK, or the meltdown of the nuclear facility in Fukushima in Japan suggest that the effect can be to entrench regressive politics. As noted above, Greens also need to be wary of the ways in which the urgency of tackling the ecological crisis can be invoked for anti-democratic ends and pursued through non-democratic means. This can take the form of overriding planning decisions (such as that of Lancashire council in the UK against fracking, overridden by national government) or speeding them up (to accelerate the adoption of nuclear power plans by invoking the urgency of transitioning to a low carbon economy, for example). Urgency can be used to trump and bypass political conflict, in a way that has been referred to as 'post-politics' (Swyngedouw 2010), and to accelerate the diffusion of controversial technologies (geoengineering, genetic engineering, negative emissions approaches) or to suspend forms of political engagement deemed incompatible with business-as-usual politics and economics on the basis that we have to 'go with the grain' of existing actors and institutions. As Stirling puts it (2015): 'urgency compels obedience'. Democracy is increasingly dubbed a 'failure' or a 'luxury' that cannot be afforded—or even queried as an 'enemy of nature'. The iconically influential environmentalist, Jim Lovelock, insists that 'democracy must be put on hold for a while' (Hickman 2010).

Greens need to be careful what they wish for. To take another example, Greens routinely support calls, such as those made at the Copenhagen climate summit, to drastically increase levels of climate aid for mitigation and adaptation, and Green party manifestos contain a commitment to increase such aid. But therein lies another dilemma. Such increases in aid to pay off ecological or carbon debts to the developing world might require substantial increases in economic activity which, in a system as tied to fossil fuels as the current global economy, implies further emissions of greenhouse gases to tackle the problem of climate change! It is rather like proposals to fund adaptation to the effects of climate change through a levy on air travel. Not a bad idea on the face of it, until you realise you are tying the access of the world's poorest people to adaptation funding to increased flying by the world's richest citizens. Such are the contradictions of modern capitalist life: the solutions proposed often intensify and worsen the problems they are meant to address because they fail to address the source of the problem.

Articulating a post-growth politics which is able to deal honestly and openly with the trade-offs that come from efforts to put welfare over output, and which actively encourages a slow-down in conventional economic activity without further immiserating the poor, will be a huge challenge. Globally, it will require striking deals on resource use which recognise the rights and needs of poorer groups to use a greater share of remaining ecological space to service their pressing development needs while ensuring that richer countries accept the greatest burden by drastically reducing their own demands upon global resources in recognition of the historical responsibility they bear for our collective predicament to date. Ideas about contraction and convergence, as sketched above, move in that direction. Domestically, it will require a major re-balancing of the economy away from financialisation, militarisation and globalisation and towards a re-centring of the economy on productive, worth-while, welfare-enhancing, employment-creating forms of work and exchange that serve the common good.

Science, in and of itself, will not be the driver. In the near term, creating winners within capitalism from shifts to a zero carbon economy is important, but that comes with other problems with which we are all too familiar. The challenge is how to address, contain and reverse 'slow violence' and how to deliberate on and develop responses to complex social crises before a situation of emergency prevails. Citizen action will be key: putting down limits to new extractivism through direct action and resistance against actions which fail to face climate reality as the Extinction Rebellion is currently trying to do (Farand 2018). It will also require a re-negotiation of social contracts within and between societies and across generations in ways that will place significant strains on today's democratic institutions.

Chapter 3

MAKING THE BEST OF CLIMATE DISASTERS: ON THE NEED FOR A LOCALISED AND *LOCALISING* RESPONSE

Rupert Read and Kristen Steele

That we carry on like this is the catastrophe.
— Walter Benjamin (1940)

The recovery of [a sense of] purpose and closeness without crisis or pressure is the great contemporary task of being human. Or perhaps the dawning era of economic and environmental disasters will solve the conundrum for us more harshly.
— Rebecca Solnit (2010)

Our changing climate demands a fundamental shift in the way that we live. Continuing down the path of economic globalisation will only create greater human suffering and environmental problems. The time to act is now. Through localisation, we can begin to rebuild our communities and ensure a healthy planet for future generations.
— Helena Norberg-Hodge (2008)

When disaster occurs, one can simply be defeated by it. Or one can use it selfishly, making things worse for others. Or, one can even make something better than what went before. In this chapter we present how the latter may be accomplished through a localised response when faced with disasters wrought by anthropogenic climate change. To illustrate, we share our contrasting personal experiences of a recent widely-known disaster and then highlight a range of examples of community responses from the field of Disaster Studies. Finally, we conclude with the practical, intellectual and creative steps needed to weather, and even make the best of, the upheaval to come.

Before that, however, it is important to note that in recent decades most disasters have resulted in a doubling down of global capitalism, which has exploited vulnerability to push privatisation and corporate control—a process that has come to be known as *disaster capitalism*. We believe there is another way, one that builds on the extraordinary ordinary natural inclination for people to come together in solidarity and creativity after a disaster. This creates an opportunity to strengthen and rebuild local economies that put the well-being of people and of the planet at the centre: a sort of *disaster localisation*, if you will.

We know now that dangerous anthropogenic climate change is not merely a matter of a slow steady increase in world temperature. Such global overheat (aka 'global warming') is real, and may well be the worst single element of anthropogenic climate change in the long term. But the more immediately noticeable effect, for most people, will be climate *chaos*. Climate chaos has already wreaked havoc in many the parts of the world—hurricanes and typhoons of unprecedented destructiveness, island nations part-submerged by rising tides, violent storms, widespread drought, unprecedented wildfires—and we are likely to see more and more of this in the years to come. In other words, the most noticeable and perhaps the most damaging impact of the over-heating that we have unleashed will be events that will manifest as disasters.

These will *be* disasters, possibly resulting in the deaths of millions. It's a terrible thing to have to foresee, even to contemplate; let alone for it to be experienced and suffered, as it will be. However, these disasters also provide opportunity for change, which if harnessed, can mitigate their impacts. Approached in this way, such disasters can open a door to a new way of doing things that actually reduces the likelihood of further disasters and increases human well-being. The gathering ominous storm clouds really could have a silver lining.

Our personal experience

Let us describe what we mean by speaking personally. Many of us are rightly afraid of the nightmare that might unfold if true disaster as a result of climate chaos affects a place like England, where we both live. Professionally, we have both worked for many years to avert such disasters by presenting an honest critique of the global economy and its climate-changing ramifications. The organisations we work with, Green House and Local Futures, are known for promoting alternative and realistic visions for a future economy that meets human needs—material, psychological and spiritual—without compromising the life support systems of the planet. On a personal level, we have, not infrequently, been in a state of some anxiety about the future if a conscious choice is not made to transform the economy.

Yet, we have found some comfort, even optimism, in coming to understand something of the intriguing academic field of 'Disaster Studies'.

Disaster Studies yields conclusions that fly wildly in the face of 'conventional wisdom' about disasters and chaos, and that meet and answer many of our (and perhaps your) anxieties about how quickly things could collapse and how dire they could get, in our fragile, complex, 'globalised' world. How does one even get to the point where one can contemplate an alternative to frantic 'prepping' and to anxiety, or indeed to the more usual response of denial?

For both of us, what happened after September 11th 2001 provides a key illustration of the different possible futures that arise from tragedy and disaster. Kristen, an American citizen originally from upstate New York, witnessed the events of that terrible day from her adopted home in the UK. From afar, she watched in dismay as widespread shock and grief was exploited for privatisation, profit-making and undermining democratic freedoms in the months that followed.

I saw a kind of immobilising fear and sadness grip people and, as they searched desperately for something concrete to hang on to, corporate America stepped in with an intense promotion of consumerism. Then President Bush in a post-9/11 speech, said: 'Our financial institutions remain strong, and the American economy will be open for business as well.' Soon there were shopping bags with US flags on them, saying 'America: Open for Business'. Basically, people were being told, 'Just go shopping. That'll fix your feelings.' In the meantime, behind the consumerist fanfare, the Bush administration was gutting democratic freedoms and pushing a war agenda with almost no initial resistance.

I tried to talk to my family and friends in the US about it, most of whom were progressive and liberal-minded, and they wouldn't hear it. I was told that criticising what the government and big business were doing to the American economy was disrespectful to the people who'd been affected by the tragedy. It was only years later, after the war in Afghanistan was launched, that many more people realised how much they'd lost in allowing that kind of economic exploitation.

But there was another side to what happened afterward, as Rupert discovered in talking with an American friend in New York City a few months after the attack.

I had recently wandered around the 'Ground Zero' ex-World-Trade-Centre site and been concerned and downhearted by the often-violent graffiti and messages left there expressing a desire for vengeance on those who had carried out the attack (and sometimes on anyone who looked like them or shared ancestry with them). I said to my friend, whose politics were close to mine: 'It must have been just awful, being in New York in those days after the attacks. I mean: with all the death, the terrible smell, the deadly pollution, the chaos, and worst of all the ferocious yells for vengeance.' His reply completely flummoxed me. He said: 'Actually, it was the one time in my life that I ever felt part of a community. It sounds strange to say it, but it was actually a happy time. People spoke to each other. Strangers helped each other—and this is New York, remember! Distinctions fell away.' And, after a pause, he said, again, 'It was truly the one time in my life when I have *ever* felt like I was really part of a *community.*'

Kristen's observations echo what we know today about disaster capitalism, as described by Naomi Klein in her book, *The Shock Doctrine* (Klein, 2007). In numerous examples, Klein points to the tendency of ruthless elite/rich elements in society to exploit disasters and to reconstruct afterward in a way that leads to more privatisation and control of the economy by large corporations. These disasters range from political chaos preceding Pinochet's takeover in Chile to Hurricane Katrina in New Orleans. In each case, social instability created a vulnerability that private interests stepped in quickly to exploit. In the short term, this stops the upsurge of civic engagement, fervour and hope that can arise in disaster. In the longer term, the social and economic consequences become more visible. For example, education was widely privatised in both Chile and New Orleans, leading to rising tuition fees and student protests in Chile, and increased inequality and exclusion of children in New Orleans.

On the other hand, what Rupert discovered in talking to his friend is an equally real example of how people can respond after a disaster, and one that offers guidance on moving away from disaster capitalism towards a localised response. Two inspiring authors have written extensively on this type of encouraging reaction: Rebecca Solnit in a *A paradise built in hell: The extraordinary communities that arise in disaster* (2010) and, earlier, Charles Fritz in 'Disasters and mental health: Therapeutic principles drawn from Disaster Studies' (Fritz,

1996). Drawing in detail on many historical and modern-day examples—from the 1906 San Francisco earthquake, the London Blitz and Hiroshima to 9/11 and Hurricane Katrina—Fritz and Solnit suggest that disaster tends to enable new communities to be born, instantly and often lastingly.

They relate how, against our expectations of chaos and panic, people in the immediate aftermath of a disaster are often remarkably calm. How they tend rapidly to develop mutual networks of support, based on need rather than on prior distinctions, whether of wealth or ethnicity. How looting, though often assumed to be inevitable, is actually rare (and how, in any case, some of what is described as looting would be much more reasonably described as the harnessing of emergency supplies; for we are talking here, often, about people suddenly desperately short of life's necessities).[1] They describe, too, how it is at least as common for people to 'converge' on the scene of a disaster, in order to help, as it is for people to flee.[2] And how all of this — against a backdrop of fear, loss, injury, death, often affecting most of the survivors directly — happens relatively spontaneously, rapidly, and even *joyfully*. How this festival of altruism is experienced by people as something they want to do, even as something that actually helps them rather than as self-sacrificing behaviour.

Post-traumatic *growth*?

Here is a contemporary account of the phenomenon from a survivor of the great San Francisco earthquake:

> Most of us since [the earthquake] have run the whole gamut of human emotions from glad to sad and back again, but underneath it all a new note is struck, a quiet bubbling joy is felt. It is that note that makes all our loss worth the while. It is the note of a millennial good fellowship... In all the grand exodus [from the most devastated areas of San Francisco]...everybody was your friend and you in turn everybody's friend. The individual, the isolated self was dead. The social self was regnant. Never even when the four walls of one's own room in a new city shall close around us again shall we sense the old lonesomeness shutting us

---

1 Sometimes, sadly, the authorities and/or the media will describe a favoured group as gathering supplies, when the same activity engaged in by a non-favoured group is called looting. Solnit, in Chapter V of her book, documents this as having happened in New Orleans, vis-à-vis whites and blacks, respectively.

2 This phenomenon of 'convergence' was particularly striking in the case of September 2001 (Solnit, 2010: 195).

off from our neighbours… And that is the sweetness and the gladness of the earthquake and the fire. Not of bravery, nor of strength, nor of a new city, but of a new inclusiveness. The joy is in the other fellow.

— Solnit (2010: 32)

We are living, nowadays, in ways that involve us in a virtually permanent absence of community. Disasters enable this to be overcome. It is important to note that, for this overcoming to take place, typically, there must be a *disaster*, not merely an accident or something bad. Fritz emphasises this point especially. He writes that disasters need to be big enough to *not* leave 'an undisturbed, intact social system' (Fritz, 1996: 21). Only if that system is disrupted sufficiently can the new forms of community emerge. 'Disaster provides an unstructured social situation that enables persons and groups to perceive the possibility of introducing desired innovations into the social system', Fritz goes on to argue (Fritz, 1996: 56). Moreover, in disaster (though not in lesser upheavals), 'many pre-existing invidious social distinctions and constraints to social mobility are removed; there is a general democratization of the social structure' (Fritz, 1996: 66).

We hear a lot today about post-traumatic stress. Fritz and others seek to teach us about something less well-known, but just as important. It is sometimes now called 'post-traumatic growth' (Solnit, 2010: 220). Solnit offers some deft reminders of relevant etymology that are worth quoting:

The word *emergency* comes from *emerge*, to rise out of, the opposite of merge, which comes from *mergere*, to be within or under a liquid, immersed, submerged. An emergency is a separation from the familiar, a sudden emergence into a new atmosphere, one that demands we ourselves rise to the occasion. … The word *disaster* comes from the Latin compound of *dis*—or away, without, and *astro*, star or planet: literally, without a star. … In some of the disasters of the twentieth century — the big northeastern blackouts in 1965 and 2003, the 1989 Loma Prieta earthquake in the San Francisco Bay Area, 2005's Hurricane Katrina on the Gulf Coast [and, in a different way, the blackout in the Blitz] — the loss of electrical power meant that the light pollution blotting out the night sky vanished. In these disaster-struck cities, people suddenly found themselves under the canopy of stars still visible in small and remote places. … the constellations of solidarity, altruism and improvisation are within most of us and reappear at these times… This is the paradise entered through hell.

— Solnit (2010: 10)

Solnit's rather beautiful thoughts here rhyme with those of Thomas Homer-Dixon (2006), who goes one further, coining a new word, 'catagenesis' (2006: 22), meaning the birth of something original from out of something disastrous. Or, as he more bluntly characterises it: 'the creative renewal of our technologies, institutions, and societies in the aftermath of breakdown' (Homer-Dixon, 2006: 268).

So, is it possible that the rising tide of disasters that climate chaos will bring could be the (re-)making of us?[3] These works by Fritz, Solnit and Homer-Dixon make evident that, when actually tested in the crucible of back-to-back disasters, it is perhaps as likely that humanity will rise to the challenge, and be transformed for the better in the process, as it is that we will shun the victims. We will likely find ourselves manifesting a truer humanity than we currently think ourselves to have, in this climate-stressed future that we are now entering. Thus, the post-normal world can offer us a gift amidst the carnage, a gift we may well, remarkably, literally make the best of.

Preparing our imaginations

How much and how well we actually realise this gift depends on our preparing the way for it. Otherwise, we may remain beholden to the other main realistic possibility — unrestrained destructive exploitation of panic and vulnerability. For as Fritz and Solnit also make clear, things often *do* go wrong in the wake of disaster, and most often as a result of (a largely delusive, but nevertheless consequential) fear of selfish, 'Hobbesian' reactions on the part of ordinary people. A striking example is New Orleans after Katrina, when African-Americans desperate for help were painted in the media as selfish villains and fuelled a violent and repressive response by government, police and the military, as well as by 'vigilantes'.

Moreover, one needs to be wary even of well-intentioned accounts that dwell on such *elite* Hobbesianism and disempower ordinary people in the process. Solnit offers such a critique of the disaster capitalism narrative: '[Klein's book] is a trenchant investigation of how economic policies benefitting elites are thrust upon people in times of crisis. But it describes those people in all the old unexamined terms and sees the aftermath of disaster as an opportunity for conquest from above rather than a contest of power whose outcome is sometimes populist or even revolutionary' (Solnit, 2010: 107).

---

3 One of us (Read) recently put this point direct to Solnit, who responded that she could indeed see how climate disasters could be harnessed in the way outlined in this chapter.

She goes on to cite Fritz's alternative analysis of people's responses to disaster: 'Fritz's first radical premise is that everyday life is already a disaster of sorts, one from which actual disaster liberates us. He points out that people suffer and die daily, though in ordinary times, they do so privately, separately. And he writes, 'The traditional contrast between "normal" and "disaster" almost always ignores or minimises these recurrent stresses of everyday life and their personal and social effects. It also ignores a historically consistent and continually growing body of political and social analyses that points to the failure of modern societies to fulfil an individual's basic human needs for community identity' (Solnit, 2010: 107).

This community identity can manifest at both the key stages of response to disaster: the *relief* stage and the *recovery* stage. In the relief stage, non-professional first responders are key,[4] especially if the disaster is sufficiently big to actually have a transformative effect. In this respect, it is important, as we noted from Fritz, above, that some climate disasters are that big, that terrible, and thus potentially that transformative. In the recovery stage, the community needs to be involved actively (and not allow itself to be treated as a passive recipient or victim, by government or business), if bottom-up transformation is to occur.

What is sometimes revealed in disaster, then, is real community identity, which fulfils our modern lack, the very opposite of what the Hobbesian script tells us will emerge. The Hobbesian script is, we should note here, among other things quite literally that, a *script*. An alarming number of books and films and TV shows suppose that disaster necessarily unleashes the worst in human beings. Consider this, again from Solnit:

> One of the more amusing recent manifestations of Hobbes came as entertainment, starting with the 2000 American television series *Survivor*... The show seemed to reference *Lord of the Flies* and other epics of savage regression and primordial competition, but merely dropping a bunch of people in a remote location and asking them to cope might have produced uneventful co-operation or unpredictable improvisation. Instead, the show's creators and directors divided the cast into teams. The teams competed with each other for rewards. Eliminating fellow members was one of the competitive games they were obliged to play to increase insecurity and drama within teams. The goal was to

---

4 Thus it is important to consider the role of institutions that already manifest a community first-responder spirit: e.g. volunteer fire services in France. These barely exist anymore in most parts of Britain, the country from which we write. A real 'big society' is needed, to start to reinstate them, as we contemplate the coming disasters.

produce a single winner rather than a surviving society, a competitive pyramid rather than a party of cooperation. Toiling for food and shelter was overshadowed by the scramble to win out in a wholly gratuitous competition based on arbitrary rules. Capitalism is based on the idea that there is not enough to go around, and the rules for *Survivor* built scarcity and competition and winners and losers into the system. These people were not in the wilderness but living under an arbitrary auto-cratic regime that might as well have been Los Angeles or London. The producers pretended we were seeing raw human nature in crisis conditions but stacked the deck carefully to produce Hobbesian behav-iour—or rather marketplace behaviour, which amounts to the same thing here.

— Solnit (2010: 93)

There are many run-of-the-mill disaster movies that have basically the same format as this. Although, interestingly, there are also many that at least posit some kind of heroic team-building as a means to community among the chaos. Then there are the stand-out cases, real art, where the whole faux-Hob-besian architecture gets overturned, such as *The Road* and *The Hunger Games*.

On the surface, *The Road*, in both the book and the film, tells an extremely bleak story, a vision of a post-apocalyptic world indeed peopled by Hobbesian monsters. The scenario in *The Road* is so grim because the author has 'manipu-lated' the conditions in his fiction, a little like how those in charge of *Survivor* manipulated the 'reality' of the contestants. In the scenario of *The Road*, some kind of ecological catastrophe has occurred so extreme that it appears that the entire biosphere is dead, except for humans. This has led to the remaining few denizens of the world devouring each other, often literally. Yet, what is often missed is the stunningly moving, redemptive ending of the story, where the dying protagonist refuses to give up on the life of his son, refuses to take him out of the world with him, and refuses to give up on the future. The son is found, after his father's death, by a family who want to take him in, and whose dog — the blessing, added to human fellowship, of a non-human other, loved even more in the absence of other non-human life-forms — evidently persuades him to say yes. An ultra-miniature paradise, literally built in hell.

The Hunger Games may well have put many off from watching/reading it by, once again, the extreme and explicitly manipulative horror of its premise. A devastated, depleted future America is held together by a rabidly authoritarian regime of the 1%, lording it over the destitute rest in particular by subjecting them to ritual combat (as a 'reality TV' show) annually. The combat takes the form of randomly selected teen children from each of the poor districts having to seek to survive hunger and cold and fight it out to the death to the last survivor in an arena rigidly controlled by their rulers for mass public entertainment. Yet, the actual story of *The Hunger Games* is of an emergent struggle against this utterly vicious system; a system that asserts, as Hobbes claimed to reveal, that only centrally administered violence and inequality can restrain the lower classes from spontaneously tearing each other apart. In the first part of the trilogy, that struggle begins with the refusal of the two last survivors to kill each other, an extraordinary act of defiance. In the second part, that spark of defiance catches fire and in the end launches an outright rebellion, when many of the participants in the new hunger games, having been chosen by virtue of surviving all the previous hunger games, tacitly refuse to kill one another and in particular strive to keep alive the girl who had initiated that defiant first act of refusal. The third part is the story of that rebellion itself as, with immense self-sacrifice, the districts rise up and finally overwhelm their oppressors.

This story explores in gripping and moving detail how human beings can be transformed for the better 'even' by an imposed disaster. We would hypothesise that it is one of the most successful box-office films of all time *because* of this.

Planned disaster localisation

So, how do we learn from all of these examples, both real and fictional? How can we avoid each new disaster leading to a more entrenched global capitalist system, with its social impacts as well as its hefty contribution to increasing carbon emissions?

First, we need to take comfort in and celebrate the stories that tell of community-building, cooperation and resilience of the human spirit. These remarkable stories and others need to be told and re-told until we have started collectively to feel that we humans are not, underneath it all, simply selfish, greedy and competitive. We, of course, have these qualities, but we are also innately kind, generous and cooperative. It is merely that the former traits are easier to exploit for profit-making in the global economy and have therefore been emphasised in advertising, media and politics. So, appreciating all our facets as humans, especially those that do not feed the consumer culture, is the first step in empowering ourselves to respond positively to disaster.

Secondly, we need a plan. One that is strategic, inclusive and looks far into the future. Otherwise, we are bound to lurch from one disaster to another with potentially cumulatively destructive outcomes. It is a new challenge, because even the field of Disaster Studies has focussed mostly on isolated, intermittent disasters. The era of climate chaos promises a ratcheting up of connected, ramifying disasters. Will there be compassion fatigue rather than a transformative creation of community and the kind of push then towards transformational adaptation that is so badly needed?

Furthermore, we also know from the empirical literature that those who survive climate disasters, though they may have gained in community, may be *less* likely to take climate mitigation seriously afterward than before. The train of thought seems to be something like 'we survived this; we can survive anything'.[5]

What these two points imply is that in order for the scenario of climate disasters being positively transformative to come to pass, systematic efforts will need to be made, along a couple of related lines:

- Widespread sharing of the kind of knowledge we have offered above, so that those affected by climate disasters, and elites planning for them, are not 'primed' by Hobbesian narratives;
- Deliberate work to cultivate empathy both for those who have survived climate disasters and for future sufferers. The analysis of climate disasters needs to become a 'school' for thinking about how to transform our societies so that they are less prone to contribute to a further worsening climate.

These two tasks may sound daunting. But, as we have already hinted, there is a 'short-cut' that goes a long way toward accomplishing what is needed.

Just as disaster capitalism has succeeded thus far by having a neoliberal paradigm constructed and promoted ahead of the disaster, those of us working for more positive futures need to have a clear vision of what we are aiming for. For instance: we need to distinguish, in our visioning, between (our response to) 'shocks' and 'stresses'. i.e. between short- and long-term. This

---

5 For discussion, see salient chapters of George Marshall's (2015) book, *Don't Even Think About It: Why We're Wired to Ignore Climate Change*. Marshall emphasises the difficulty that the last thing people want to do after most climate disasters is to talk about something controversial, such as how the disaster they have been through was climate change-driven. But this is changing: in countries like the Philippines, wracked by unprecedented typhoons, impatience against any remaining climate denialism has boiled over, and people are no longer prepared to wait before talking about the climatic aspect of what they are suffering (see e.g. https://www.theguardian.com/environment/2014/apr/01/yeb-sano-typhoon-haiyan-un-climate-talks). Furthermore, 'attribution science' is improving rapidly (see http://threeworlds.campaignstrategy.org/?p=2069 on this); it is getting easier for scientists to pinpoint how disasters are indeed climatically-induced, and even to attribute responsibility to individual 'carbon major' corporations.

chapter concerns mainly the former, i.e. shocks, or disasters, and seeks to develop a new 'counter doctrine' that will give us a chance of seizing collective advantage via 'catagenesis'. Responding effectively to stresses is more challenging, because there is no obvious tipping-point moment,[6] and because we become subject to 'shifting baseline syndrome', gradually accepting worse and worse realities as 'normal'. The terrible thing about climate change is that it increases *both*. That is, it gradually causes more and more stress, while also increasing the number and severity of shocks.

So, which assumptions and paradigms do we need to lose and gain, if we are to be able not only to enter but to thrive in the paradises which Solnit has suggested are potentially open to us? How do we make the most of disaster and even breakdown, ensuring that this does not become simply complete collapse? How can we find a way forward that can take advantage of the opportunity unleashed by shocks, while starting also to reduce stresses?

We believe a solution lies in *localisation*. This encapsulates the many 'theories', initiatives and actions needed to create a new system that provides both human and planetary well-being. At its core, localisation is about reconnection. Economically, it means scaling down big business, making businesses accountable to the people they employ and the people they serve. It means shrinking the distances between producers and consumers, which is empowering and helpful on many levels. For one, the carbon emissions from long distance trade would be reduced. Consumers would also know more about how products were produced and who made them, enabling them to make choices that support fair labour conditions and sustainable production practices. Internationally, localisation means restructuring regulations, such as trade treaties, that benefit footloose corporations and bankrupt communities, sometimes whole nations. It also means shifting the balance of political power from multinational corporations to citizens and reviving true democratic process.

Localisation is the building of community, in a virtuous circle with whatever strands of community already exist. It is the production of empathy in real circumstances of contact and of co-creation.

Localisation thus encourages the most important resilience of all: a resilience of people and of communities, not merely a resilience of things (e.g. flood protection). Governments tend to focus on resilience of things. This is typically what is meant also by 'adaptation' to human-triggered climate change. But we need a *transformational* adaptation, bold enough to rise to the challenge of the coming climate disasters; one that is based solidly in

---

6 Although that is not as fatal a difficulty as is often assumed. See the discussion of how humans might yet become 'wise frogs', unwilling to be gradually boiled to death, here: https://rupertread.net/writings/2017/climate-change-white-swan.

understanding that what is needed above all when disaster strikes is solidarity; as well as one that calls upon government to assist these great changes, not to block or minimise them.

What deliberate fostering of localisation entails in practical terms at the level of government is widespread policy change. For example, in many countries, employers currently have to pay extra tax on each one of their employees. By contrast, they can get subsidies and tax breaks for using more energy and investing in more people-replacing machinery. Switching taxes like these around so that energy and machinery are taxed and labour is not would incentivise hiring more people and using less energy—addressing unemployment and climate change simultaneously.

Such economic and political changes provide a firm foundation for rebuilding strong, harmonious communities. When the economy is at a human scale, we connect with others directly to meet our material needs. A great example is the localisation of food and farming, a movement that has swept around the globe in the last couple of decades. Not only do localised food systems lower food miles and provide healthy food, grown with earth-friendly practices, they also encourage community building. Multiple studies have shown that farmers markets—the quintessential embodiment of the local food movement—appear to facilitate social interaction. Even if people spend the same amount of time shopping as they would in a supermarket, they have many more conversations while doing so, both with producers and other consumers. The cut-throat competitiveness of capitalism falls away between producers as well as they get to know each other, have regular contact and begin to cooperate and complement each other's products rather than compete for the same narrow market niche.

A beautiful example of this process occurred in New Orleans after Hurricane Katrina. In the 1980s, there was a burgeoning community gardens movement in the city, making use of the many vacant lots. However, as the 1990s wore on, and the city became a hot destination for global capital, many of these lots were taken over by new developments. Local people were pushed out, and by the time the hurricane hit, the poorer areas of New Orleans were considered food deserts. People in these areas mainly relied on local convenience stores, with their fried and processed fare, for their regular diets.

However, that changed in 2005 with Hurricane Katrina. Such was its destructiveness that the number of vacant lots tripled overnight. Over the next few years, people and organisations began to move into these areas and create abundance from desolation. Vegetable gardens, fruit orchards and apiaries sprung up around the city. Some of these are privately owned, many are community managed. Some are designed for generating income and some are simply to provide nutritious local food for local people. Within five years, there were more than 100 farms and gardens in the city. Many people and organisations, like the New Orleans Food & Farm Network (NOFFN), now continue to build and strengthen the local food economy by creating food maps, training growers, setting up farmers markets, and connecting growers, processors and consumers.

Similarly, in the wake of the widespread devastation from two recent hurricanes in Puerto Rico, many locals have joined together to reject the kind of disaster capitalism that has kept the island in crippling debt for decades. As Elizabeth Yeampierre and Naomi Klein (2017) reported, after the hurricanes, Puerto Ricans were already on 'the lookout for how these shocks would be exploited for private gain. The destroyed electricity grid would be seized upon to argue that the whole system should be privatised, while the destroyed homes would be the opening to auction off more land for golf courses and vacation homes.' In a pre-emptive response, a campaign for a 'just recovery' was launched. A coalition of groups, including Our Power and Organización Boricuá, is now working on a multi-pronged plan to push for a recovery that includes debt relief and prioritises equity, food sovereignty and environmental justice. As Yeampierre and Klein put it: 'Puerto Ricans are hard to shock, but the island may be on the verge of shocking the world by seizing a crisis of unimaginable hardship to forge an inspiring new model of economic development'.

Such examples give an idea of the lasting future we can create out of disaster by applying the principles of localisation. But it is clear that localisation needs to happen at both the grassroots and policy levels. At the grassroots, we're seeing tremendous growth of initiatives around the world. Besides local food, there are local banking initiatives, localised education, localised medicine and so on.

Further examples feature elsewhere in this book. For instance, the lessons from Lancaster after flooding in recent years as presented in Anne Chapman's chapter. They are sobering but also encouraging. The way that the sudden removal of electricity forced people to talk to one another, to help one another, to congregate and communicate, shows us something about what we can gain when we lose technologies that we are accustomed to, and what we can lose when we gain those technologies. What is needed now is to make the connection clearly with climate. And then to enact more rapidly the programme of transformational adaptation that is called for.

However, short of a few regulatory changes at the municipal level in a handful of countries, almost nothing has happened to shift national and international policy towards localisation.

Part of the impediment to such change is the reliance on economic growth as an indicator of well-being. But GDP (Gross Domestic Product) only measures money changing hands, which means that money spent on oil spills, car accidents, cancer treatment and so on adds to the positive side of the balance sheet. Global trade, with its complicated supply chain, corporate control and massive use of fossil fuels, is a great contributor to GDP. As Green House and Local Futures have been arguing for years, there is a dire need to abandon the harmful growth mentality and seek to create a post-growth future. Such a future will be based in localisation.

The growth mindset spans all mainstream political parties, and despite multiple other metrics being proposed (such as the Happy Planet Index and the Genuine Progress Index), the world economy is still on a path of growth-at-all-costs. The costs include impending climate chaos and the disasters we are already seeing and that will ensue. But, as we pointed out earlier, if we can harness the natural impulses of people to come together to work together, there is great hope that we can emerge better and stronger than before. Thus, 'post-growth' and 'pro-localisation' go hand in hand.[7]

Disaster localisation is in any case coming, like it or not. It will come eventually in a forced, difficult and uncomfortable way, if some variant of business-as-usual leads our civilisation to collapse,[8] as now seems quite likely. In that scenario, eventually people will only be able to respond locally to disasters. This is the ugly potential face of climate reality.

Or, a more vital, sustainable and pleasant kind of localisation can be created in the manner argued for in this chapter. This will be challenging to bring about in the face of massively powerful opposing political and economic forces and assumptions. But it is so preferable to the alternative that it is worth fighting for, and hard. What we hope and suggest is that climate disasters, if we together manage to foment a way of receiving and understanding and responding to them along the lines that we have set out here, can themselves be the catalysts for this change.

What we need now, therefore, is to rally around a vision of localisation that is inclusive, diverse—*and ready to be implemented in times of crisis*. This will then prevent the downwards slope of economic, social and environmental exploitation we have seen after so many previous disasters. Having such a multifaceted and adaptable vision in place offers a blueprint for constructively

---

7 For more on the convergence of post-growth and localisation arguments see the pamphlet Post-Growth Localisation (Norberg-Hodge and Read, 2016).
8 See Rupert Read's article: http://www.truthandpower.com/rupert-read-some-thoughts-on-civilisational-succession/.

engaging people's natural post-disaster impulses and creating a better future—the essence of 'disaster localisation'.

We can see clearly the outlines of a gift that climate disasters can bring in their wake, if we are ready to accept it. A gift that would stop us having to look for silver linings as we descend into collapse and enable us to genuinely head off that collapse—an *unleashing* of *humanity*, of decency, of care. Because these disasters show the future of the business-as-usual scenario. They show it in its horror. But they also show us the riposte.

The scale of the climate crisis calls for systemic change. The systemic change needs to involve a deliberate fostering of community and of low-impact living. This is exactly what the transformational programme of 'localisation' offers. And thus, even in the darkness of this time, we dare to hope for a better future. For the darkness of disasters can nourish the extraordinary brightness of the 'silver linings' we have outlined above. Where the danger lies, there also the saving power resides,[9] ready to flourish.[10]

---

9 'Wo aber Gefahr ist, wächst / Das Rettende auch' (Hölderlin, 1951:165).
10 Deep thanks to Green House and Local Futures colleagues for comments that have enriched this chapter. This chapter was born in the recent debate between Rupert Read and John Foster in Global Discourse (Read, 2017) and Helena Norberg-Hodge's pioneering work on local economies shaped the vision of localisation presented within.

Part II: Systems

Chapter 4

LINKING CITIES AND THE CLIMATE: IS URBANISATION INEVITABLE?

Jonathan Essex

The fight for beauty is not blind opposition to progress, but opposition to blind progress.
— Sierra Club Motto, David Brower (Robinson, 2001)

Introduction

The links between cities, urbanisation, globalisation and climate change run deep. Continuing globalisation and urbanisation are reducing the resilience and sustainability of our built environment in the face of climate changes.

Urbanisation is forecast to increase city populations in the global South particularly, in some of the most climate-vulnerable regions of the planet. The process of making these cities bigger has huge carbon emissions attached—both through the carbon embodied in construction, and other products and food imported into cities, which are often excluded from 'place-based' estimates of carbon emissions. Yet talk of 'sustainable development' and a 'circular economy' have not changed the overall way in which we are increasing the scale of our construction, production and consumption of resources and energy worldwide. It appears that only lip service is being paid to the realities of the scale of what climate change will do, and what avoiding its worst effects will require.

While urbanisation, globalisation, and significant infrastructure (especially transport infrastructure) investment continue largely unabated, the visible impacts of climate change, such as the melting of the Antarctic, are accelerating. The climatic imperative to change the pattern of our settlements and development is not reflected in strategic planning and economics. Continuing these 'business-as-usual' trends will most likely snowball impacts from isolated local and regional disasters to an avalanche of negative interactions with the global economy, leading to a system failure.

This chapter first explores current urbanisation trends and how these impact on climate vulnerability and carbon emissions. It then explores current economic and investment strategies before addressing what a different way forward might entail.

Climate change calls for a different urban form

Continued urbanisation and its development challenges

Urbanisation is predicted to increase the proportion of the world's population living in cities from over 50% today to 75% by 2050, with over 95% of this growth occurring in developing countries (Atkins, UCL and DFID, 2015). In many countries, there is strong rural-urban migration, in part due to climate and environmental pressures and changing rural livelihoods, as land is taken for more commercial, export-led agriculture, and in part due to the pull of cities, primarily for work. This is to some extent replicating the urbanisation that has already taken place in many more developed nations.

Urbanisation is increasing the size of not just of the largest, but of *all* cities. Although more and more people are living in mega-cities, urbanisation is most rapid in smaller cities with 500,000 or fewer residents, which constitute half the world's urban population (Cohen, 2006). In addition, globalisation has, in effect, added an extra tier of specialised cities on top of the network of capital cities, which is extending the traditional size distribution of cities (Evans, 1972). This phenomenon of 'global cities'—how cities interface and drive globalisation—has been explored by Saskia Sassen (2012).

Global trade is locking in much of this new urban expansion in the world's most climate-vulnerable cities. This is because ports and urban areas often combine as transport hubs. Expansion of cities in low-lying coastal zones could represent a combined population of 1 billion by 2060 (Cohen, 2006).

The location and scale of city expansion are increasing both carbon emissions and climate vulnerability, as well as causing wider social and environmental consequences. As the urban half of the world's population accounts for some 75% of carbon emissions this reflects the energy and resource demands and pollution impacts which cities place on areas outside their boundaries.

And the trend towards larger and larger cities requires increasingly global-scale supply chains to feed their appetites for food and energy, raw materials and consumer goods. The huge distances involved reduce the capacity of cities to survive without external links, increasing their vulnerability.

The United Nations Development Programme (UNDP) highlights the fact that the people with the highest vulnerability to climate impacts are the 40% of urban residents—2 billion by 2030—who live in 'informal settlements', or slum areas (Klugman, 2009). These include many who have migrated from rural areas where they may not have had to pay for their home, food and renewable resources (as these are partly outside the formal economy), so while the new urban poor may be richer on paper than the rural poor, they often live in poorer living conditions in more (climate) vulnerable locations. There is often a local tension between trying to improve existing urban areas and their rapid expansion often through new informal settlements. Klugman refers to the 'closed city' policy of Jakarta, Indonesia and evictions and slum clearances in Dhaka, such as the clearance of 60,000 peoples' homes in 2007, as an example of how the most vulnerable people are not the subject of long-term city plans. Addressing this growing vulnerability requires different national and global, as opposed to just city-scale, planning and policies which restrict, remove or fail to recognise people living in informal settlements. And that is before the impact of more rapid sea-level rise due to polar and glacial ice melt is reflected in decision-making going forward.

Finally, as highlighted above, global carbon emissions must peak far sooner than world population is predicted to peak (currently predicted to keep on rising—to around 9.8 billion by 2050, and 11.2 billion by 2100: UN DESA, 2017). Current rates of urbanisation are projected to increase urban populations to an estimated 6.1 billion by 2050, with demand for food and water rising in parallel. New *et al* (2011) highlight that faster temperature rises due to climate change by the 2050s or 2060s could coincide with peak population demands for food and water, magnifying consequences. In contrast, far stronger climate mitigation now will lower overall temperature rises and delay more extreme climate impacts until demand for food and water has (potentially) peaked or begun to decline alongside global population.

In response to these challenges, many are calling for changes in urban form—for a combination of improved city resilience, investment in 'smart' technologies, plans for cities to be spatially compact, economically inclusive and lower-carbon, and so on. This chapter is not an analysis of how these can together improve *existing* urban areas—but of how *continued urbanisation* links to climate change, and what the alternative to this might be.

How urbanisation increases climate vulnerability and carbon emissions

Predictions (see section above) note that the world's population is continuing to urbanise. This is increasing carbon emissions and vulnerability through three separate, yet interconnected, factors: increasing the scale of the built environment, how this increases production and consumption of goods, and how these interact with climate change. These are explored in that order below: firstly, the scale of construction, and where these growing cities are located; secondly, the associated impact on the scale of production and consumption; and finally, how these are affected by (and impact) climate change.

Firstly, if 75% of people will live in cities by 2050, and global population increases to at least nine billion, as predicted, then this would double the number of people living in urban areas, from around 3.5 billion in 2015 to more than 7 billion by 2050. Assuming national infrastructure investment (e.g. water supply systems, transport networks and power supply and distribution) is proportional to city growth, as opposed to wider population, that means almost a doubling of physical infrastructure worldwide.

And this is before any consideration is given to continued physical development and building work in countries that have already urbanised. Continued investment in increasing the scale of existing infrastructure and buildings is what underpins continued economic growth in many countries—investment in such 'fixed capital asset investment' is not just a feature of urbanisation but a feature of 'sustained' economic growth in many countries (Essex, 2014).

This implies a vastly increased demand for resources: timber, concrete, steel, brick, plastics and so on. Such a scale of urbanisation until 2050 implies new construction equal to all the existing urban built environment. This is reflected in predictions by economists—some suggesting that an additional $90 trillion investment in infrastructure will occur by 2030, which equates to more than everything that exists already (Global Commission on the Economy and Climate, 2018).

Assuming buildings continue to be built as they are today, then the amount of 'embodied carbon' (even before subsequent 'in use' energy) could be enough to cause runaway climate change—even if the subsequent living is zero carbon, with a circular economy and so on. For example, over 50% of Shanghai's carbon emissions in the early 2000s was due to the construction industry alone (Baoxing, 2007): the urbanisation process is both creating a lot of carbon emissions directly and locking in significant further increased carbon emissions in the future. This would worsen the plight of many moving to cities now and is placing cities at the fulcrum of a vicious circle between increasing greenhouse gas emissions and climate vulnerability.

Secondly, the scale of energy and resource use in cities should be considered. A number of city-wide studies explore trends in increased in-use energy demand in cities due to climate change (Hunt and Watkiss, 2011). In China, escalating summer peak electricity use is attributed to air-conditioning in cities (Baoxing, 2007). A 1°C rise in temperature in Shanghai is estimated to have led to a 3.67 Gigawatt-hour increase in daily electricity consumption (Li, 2013). The total energy use should consider not just that consumed in a city itself, but how this draws resources from further afield. For example, London's overall consumption emissions are around twice those produced directly within the city' physical limits (BioRegional, 2009; based on data from 2004–7).

Thirdly, the average size of cities is increasing. Clearly, cities will be lower carbon if they are compact rather than sprawling in nature (Global Commission on the Economy and Climate, 2014 and the Electric Wallpaper Company, 2004) but their aggregate carbon and resource impacts will still tend to increase as they get larger. The ecological footprint of London was found to be twice the area of the UK, with the average ecological impact of a Londoner higher than the average of those living across the rest of the UK (Best Foot Forward, 2002).

Economic structures, including globalisation, allow cities to grow far beyond the carrying capacity of their rural hinterland, through trade links. Cities are becoming more vulnerable as they become increasingly reliant on global supply chains. This vulnerability was highlighted by Andrew Simms, reflecting on the three-day blockade of fuel supply depots in the UK in the summer of 2000 (Simms, 2008). City states (such as Gaza) and city-scale refugee camps (such as the five camps near Dadaab in northern Kenya) are particularly vulnerable, due to their higher dependence on food and resources from outside. This could be an issue as climate impacts threaten the scale and reliability of global food supplies, whilst climate change itself is increasingly a factor in the displacement of people within and between countries (World Bank, 2018).

Finally, whilst there is significant discussion about the need to divest from what will become 'stranded' fossil fuel assets, the types of infrastructure and places that rely on high levels of fossil fuels to supply energy should also be explored: some major cities and their global supply chains will become unviable. We should question whether it is sensible to continue to invest in climate-vulnerable mega-cities worldwide. This is explored in the next section.

Climate change is already increasing the vulnerability of cities

A disproportionate number of major cities are in vulnerable low-lying areas. Climate effects are converging to produce multiple impacts on cities, particularly those continuing to expand in the developing world (Atkins, UCL and DFID, 2015). The top five cities in terms of population exposure are all in Asia: Kolkata and Mumbai in India, Dhaka in Bangladesh, Guangzhou in China and Ho Chi Minh City in Vietnam.

Rather than policies to avoid expanding cities in such vulnerable locations, however, the current approach remains an incremental response which prioritises improving the (short-term) resilience of cities and other settlements to climate change impacts, rather than transforming the relationship between cities and climate change. City-wide visions that do combine environmental sustainability and resilience still rarely challenge the notion that urbanisation should continue, which would conflict with wider politics and economics. This 'better business-as-usual' approach assumes that it will always be possible to address climate risk incrementally—and that this is a credible alternative to directing development away from cities, which have economic importance.

Cities which are already highly vulnerable to climate-related disasters continue to develop. Examples include Dhaka (Bangladesh), Dublin (Ireland), Karachi (Pakistan), Bangkok (Thailand) and London (UK). These large economically important cities remain main focal points for investment and growth, even though they are already highly vulnerable to climate impacts including flooding, drought and sea-level rise. This may turn out to be a foolhardy approach. A good case study might be the medieval town of Dunwich, once the tenth largest city in England but finally lost to the sea after a series of devastating storms in the 13th and 14th centuries.

What would be the impact of a series of mega-storms today? And what if they were to affect many global cities simultaneously? By the 2070s there is a 99.9% chance of at least one city 'being affected by a 1:100 return period event' every five years—which equates to recurring city-scale disasters at the global scale (Hanson et al, 2011). What if business-as-usual climate emissions led to progressive failure of the West Antarctica Ice Sheet and around 7m of sea level rise in a relatively short time? This surely would have a major impact on trade and global supply chains, forcing a radical shift in the pathway for development—albeit a shift that would be possibly too late to avoid dangerous climate change. Might an appreciation of the true scale of these future risks galvanise a sufficiently radical shift in our approach to economics, planning and development today?

One typical aspect of current export-led economies, with development tied to globalisation, is the need to develop key transport infrastructure: principally national road and rail networks, airports and ports. Yet although 13 of the 20 largest cities worldwide are port cities, there has been little preparation for such future climate events (Hanson et al, 2011). Meanwhile, the increasingly airport-led focal points of globalisation (Freestone, 2009), combined with the continued scale of global shipping's impacts could lead carbon emissions to spiral further out of control.

This lack of foresight is also reflected in international aviation and shipping still being excluded from international carbon agreements, 25 years after the international climate negotiations started in earnest. This leaves the key relationship between urbanisation, globalisation and climate change excluded from strategic economic planning. Climate scientists, Kevin Anderson and Alice Bows-Larkin (2012) describe the challenge for shipping and climate change as requiring a Scharnow turn (that is a man-overboard rescue turn, passing a point previously passed through). But is climate change a consideration in international trade deals, establishment of export processing zones or customs arrangements? Together, these questions point to a need to re-examine seriously the climate consequences of binding together the world's cities through the current scale of international trade.

Enhancing 'disaster resilience': prioritising short-term risk over longer-term sustainability?

One response to the increasing frequency, severity and unpredictability of disasters has been the growth of a new discipline: *disaster risk reduction*. The potential strategies for improving resilience are, in theory, many and varied. The approach taken varies from financial (disaster risk insurance) to focusing on improving the physical resilience of communities or assets, including through better maintenance, strengthening community resilience and that of overall infrastructure systems (including strengthened participation, early warning systems, disaster preparedness). However, the tendency is still predominantly to increase the resilience of settlements and infrastructure where they already are, in ways that do not challenge the direction of future development.

Firstly, risk insurance puts aside money in case a disaster hits but does not automatically lead to planning or actions to reduce the likelihood of disaster. This approach tends to sit alongside continued economic development, changing the direction of development only incrementally, if at all.

Secondly, improving the resilience of the built environment tends to invest further in what is already built, often reinforcing existing infrastructures and behaviours, using the same materials and solutions. For example, using common building materials such as reinforced concrete, welded steel and bricks bonded together with Ordinary Portland Cement (which is stronger than traditional lime mortar) in disaster-prone areas limits how much can be reconstructed after a disaster. Similarly, disaster risk is often reduced by increasing the size of existing bridges or sea defences, with wider economic activity largely unchanged.

Such an approach might improve resilience in the short-term but could be catastrophic at some point in the future. For example, the current practice of adapting to climate change in Bangladesh includes progressively raising road embankments for national roads to sit 0.5m above the historic maximum flood level. This means that roads running east-west increasingly constrain flood waters in the rivers flowing north-south. Similarly, raising sea walls can protect inland areas from coastal flooding (e.g. in Guyana, the Netherlands, eastern England), which might reduce the likelihood of extreme events in the short term but may not suffice in the longer term.

Kolkata, India and elsewhere: the danger of optimism and incremental change

One increasingly common approach is to integrate climate resilience into planning, at the city scale. The approach currently being taken by Kolkata, India—the city considered the most vulnerable to climate change globally by 2070 (Doig and Ware, 2016)—is typical. The recent Road Map for a Low Carbon and Climate Resilient Kolkata (Price Waterhouse Coopers, 2015) is based on a predicted sea level rise of 0.27m by 2050, which is based in turn on science already a decade old. Even on this basis, the annual expected climate-induced flood damage in Kolkata could be US$5 billion by 2050 (Dasgupta et al, 2013 based on modelling by IPCC, 2007). But what would be the human cost if the true extent of already locked-in sea level rises were factored into city planning? The last major floods in Kolkata in 1978 put much of the city several metres under water—what might happen in the future?

Kolkata is not unique in basing its planning on decade-old data. Most policy is based on IPCC reports, which lags behind peer-reviewed papers, which lag behind the most recent research. This means there is still a lack of sufficiently long-term, realistic planning, and little to redirect urbanisation away from making mega-cities even larger, increasingly in climate vulnerable locations.

Bangladesh is among a growing number of countries (including many small island states) where the impacts of climate change on physical infrastructure and people's livelihoods are already critical and are reflected in increasing vulnerability both in rural areas and major urban centres, as well as through current urbanisation.

Climate change appears to be accelerating migration to informal slum areas in the capital city, Dhaka. While it is clear that migration underpins urbanisation, which has led to growth in the Bangladeshi economy, it is not clear what has driven this migration—is it the *pull* of the successful garment sector or the *push* of climate change?

Some suggest that Bangladesh's 'development surprise' is driven by the successful ready-made garment industry, which has expanded at over 15% a year since 1990, driven by rural-urban migration as a means of delivering upward mobility for economic migrants (Mahmud, Ahmed and Mahajan. 2008). Around 70% of the 4 million garment workers on whom Bangladesh's garment trade depends are migrant women. However, some challenge this causality and argue instead that it is climate displacement and not the pull of cities that drives migration into the city slums of Rajshahi, Khulna, Chittagong and Dhaka (Shadsuddoha et al, 2012; Kniveton, Martin and Rowhani, 2013). Sassen (2014) goes further, tagging these migration trends not so much as positive choices but as being about the *expulsion* of people from vulnerable rural areas, due not least to climate impacts.

The process of displacement is often triggered by major disasters. Displacement Solutions in Bangladesh (2012) recorded that of the 60,000 people internally displaced by Cyclone Alia in 2009, 25,000 were unable to return to their houses and, with little access to new land, tried to live on a 25km long, 2m high and 3–4m wide embankment. For these people, a cyclone shelter is insufficient: they need somewhere to safely build a house and secure a livelihood. Many instead chose to leave a precarious rural existence and head to cities. Research by Shamsuddoha and Chowdhury (2009) highlighted that even a 1–2°C increase in temperature (which could still equate to *avoiding* dangerous climate change) would force physical dislocation of more than 35 million people in Bangladesh alone.

But the current main destinations for Bangladeshi migrants, Khulna and Dhaka, are both close to sea level. Expansion of the built environment in Dhaka into its lowest lying area, which was previously set aside as flood plain, will increase future disaster risk (Haque, Grafakos and Huijsman; 2012). Migration to these cities which, like 90% of Bangladesh's land area, are less than 10m above mean sea level, will increase overall vulnerability. The last storm surge over 10m high in Bangladesh was in 1970 and killed around 300,000 people.

Conclusions for this section

Bangladesh's continued rural-urban migration to fuel the garment trade highlights the risk of a short-term economic strategy. Similarly, the inadequate sea level rise projection used in Kolkata's urban planning highlights the risk of focusing on short-term resilience at the expense of long-term sustainability.

Reduced agricultural production due to climate change could mean migration from heat- and drought-stressed areas in Africa and coastal areas at risk of flooding, as explored above. But what if the people moving in this growing mass migration find themselves increasingly hemmed into large, resource-insecure and climate-vulnerable cities? One possible future will see Bangladesh's major cities (and those of many other low-lying countries and small-island states) increasingly functioning as export processing zones, where internally displaced people are exploited as cheap labour by financial interests mainly based elsewhere in a globalised world.

So, while people can, in the short term, relocate from rural to urban areas, or across borders, as crops fail and deserts advance, this could escalate the number of fragile cities and states and hence increase conflicts in the longer term. The direct consequences of climate change (e.g. flood, storms, extreme heat) also have knock-on effects, notably food and water insecurity, turning natural disasters into complex emergencies. Climate change is already having an impact on conflict, security and fragility, for example in Darfur (Sudan) and in across the Sahel. Climate has also been identified as playing a complicating role in more recent conflicts following the Arab Spring, though no conflict has a single explanatory factor (Peters and Vivekananda, 2014).

Responding to climate change from an adaptation point of view alone tends to be short-term, and insufficient, as unless the world collectively eliminates global carbon emissions there will come a point when adaptation is woefully inadequate and the resource capacity to sustain humanity will plummet. Relocation alongside competition for scarce resources presents an undesirable future for many.

However, an alternative still exists, and this is *sufficient* adaptation *and* mitigation of climate impacts *together*—and for this to apply to strategic planning beyond the scale of the city. Alternative approaches to reduce the exposure of a city to climate impacts could include upstream catchment management or coastal mangrove regeneration—widening the 'area of influence' to be considered when exploring alternatives. The question of how much more inclusive decision-making should be is explored in the next section.

Current economic planning encourages urbanisation

Urbanisation: driven by infrastructure-led, not climate-led, investment

Most infrastructure investment decisions do not prioritise addressing the gap between climate science and current actions: instead investment is often proposed to meet a so-called 'infrastructure gap'. This means continuing to scale up infrastructure investment overall to enable continued urbanisation and globalisation of the economy. For example, the Programme for Infrastructure Development in Africa (PIDA)'s priority action plan is valued at

$65–70b, over 50 times the scale of the African Development Bank's Pilot Programme for Climate Resilience. Similarly, the UK's National Infrastructure Commission's plan (HM Treasury, 2014) focuses on the so-called 'gap' in infrastructure investment. They do this because capital assets are crucial to sustain increases in production and consumption levels, thereby sustaining economic growth. But investment in long-distance transport infrastructure and fossil fuel energy extraction capacity directly increases carbon emissions, thereby reducing long-term climate resilience globally (Essex, 2014). This approach sees climate resilience as a constraint on investment. Such planning and investment decisions are not governed by the UK's national five-year carbon budgets, even when they directly increase carbon emissions. Instead, the UK has a strategy to ensure that infrastructure is resilient for a changing climate, an approach which, if combined with inadequate climate mitigation globally, will underestimate the scale of climate adaptation required in the long-term (UK Government, 2011).

In the UK, a National Infrastructure Plan and Industrial Strategy now inform how infrastructure can grow the scale of the UK economy (HMG 2014, 2017). The former contains over 100 large infrastructure projects, including adding third runway at the UK's busiest airport, totalling £377 billion of public and private investment. Alongside such large schemes investment in the UK energy sector is shifting to smaller, decentralised generation, but subsidies are encouraging this sector to speculate on exploring for unconventional oil and gas extraction rather than investing in large-scale deployment of onshore renewables. Although there is now a shift to electric car sales (up from 3,500 registrations in 2013 to 150,000 by May 2018), this is not matched by a modal shift to public or active transport, or by any reduction in freight impact.

In contrast, research for the Organisation for Economic Co-operation and Development (OECD) argued that the focus of investment should extend beyond the climate resilience *of* infrastructure to investing in infrastructure *for* low-carbon and climate-resilient development (Corfee-Morlot et al, 2012). This research states that we have a unique opportunity to shift investments towards low-carbon and climate-resilient infrastructure, to finance the transition to a low-carbon, climate resilient economy. Similarly, research by Hagemann et al (2011) states 'there is a need to bridge the gap between tools for adaptation, mitigation and development (...) to integrate adaptation, mitigation and development at different levels and across geographic regions.' Therefore, zero carbon and climate resilience should be embedded when investment choices are selected, rather than added to traditional investment plans that scale up existing, particularly transport, infrastructure.

Estimates for the scale of 'green' investment requirements are still substantial. In 2010, Ernst and Young estimated that the UK needs to invest some £450 billion in low-carbon development until 2025, split evenly between £225 billion in the energy 'supply side' and £225 billion in the energy efficiency 'demand side' (TUC, 2011). The Global Commission on the Economy and Climate (2014) estimated that $93 trillion of infrastructure investment is needed in transport, energy and water systems, much of it in cities, to meet global infrastructure needs in the next 15 years, while also ensuring the transition to a low-carbon economy. This reflects the mainstream focus on continued growth alongside efforts to make existing cities more sustainable, inclusive and lower-carbon. Similarly, the OECD (2015) notes that to meet the 2°C climate target, energy efficiency investment needs to increase eight-fold by 2035 compared with 2013 levels. This suggests that investment in low-carbon and climate-resilient infrastructure, as opposed to overall 'infrastructure' spend, should be accelerated.

But this is not happening. The tendency worldwide is to focus on the infrastructure-gap, to include resilience in this, but to limit the focus on mitigation to low-carbon energy and not to consider the lock-in of carbon emissions from infrastructure investment in general.

Economics first, climate change (and people) later or absent

The general distortion of investment detailed in the previous section is occurring because climate change is:

- firstly, not the priority of most strategic economic planning; and
- secondly, often under-estimated or excluded from economic models that underpin such investment decisions.

Nicholas Stern (2013) contended that economic models further limit risks by assuming *'only modest damages from climate change and narrow distributions of risk,'* and he recommended that new models are required. The most commonly used economic modelling underestimates economic impacts due to unrealistic assumptions. Dietz and Stern (2014) highlight how the widely-used Nordhaus model of climate cost-benefit predicts that 18°C of warming (yes, that is what it predicts!) is required to reduce global output by 50%—a massive underestimation of explicit and large climate risks—as 1.5–2°C warming is considered extremely dangerous by climate scientists.

Another problem is that investment rarely considers the full carbon emissions and climate impacts over the whole lifespan that infrastructure is predicted to last, *and* how long cities are expected to remain in their current location, but instead focuses on a 'return on investment' period. The latter tends to be incredibly short. For example, the World Bank's typical benchmark return period for transport infrastructure investment decisions is a 12% internal rate of return. This means that globally significantly infrastructure investment is typically designed to 'pay back' financially in 6–7 years—so climate impacts are rarely a governing factor. Similarly, investment to create new 'economic corridors' tends to connect cities, rather than cities to their rural hinterlands (for example in UK, Europe, Nigeria and Asia: HM Treasury, 2018; EU, 2015; World Bank, 2018; Brunner, 2013). Upfront investment in infrastructure is then justified in terms of economic returns from rising land-price values realised through subsequent real estate investment and continued urbanisation. Such projects are not constrained by climate impacts—short or long-term. Economic discount rates make matters worse. For example, cost-benefit analysis using the net-present-value approach that accounts for climate impacts can show *increased* viability of many projects, which cannot even be justified based on current weather data (such as in Kolkata: Dasgupta et al, 2013).

This approach risks not just failing to address the challenge posed by climate change, but locking-in the status quo. The 2014 Intergovernmental Panel on Climate Change (IPCC) working group 3 chapter on human settlements, infrastructure and spatial planning (Seto et al, 2014) noted that:

> The global expansion of infrastructure used to support urbanisation is a key driver of [greenhouse gas] emissions across multiple sectors. Due to the high capital costs, increasing returns, and network externalities related to infrastructures that provide fundamental services to cities, emissions associated with infrastructure systems are particularly prone to lock-in ... especially for energy and transportation infrastructure.

Avoiding this, requires *all* strategic investment plans to prioritise *both* climate resilience *and* low/zero carbon investment—together. This requires an entirely different pattern of investment, and subsequent development.

The reality is that current investment trends amplify the risks of runaway climate change. This is rarely reflected in the economic models and national-level spatial planning that guide economic investment decisions. If this was the case it would be transformative—as investment would then be limited to that which explicitly *reduces* carbon emissions. As these numbers are not translated into rules-of-thumb or modelling that informs mainstream economic decision-making (e.g. level of concrete, steel and brick use in national carbon reduction targets; impact of 1–2m sea level rise on combined storm surge and sea level rise flood risk), climate considerations tend to be entirely excluded from decision making, or at best limited to carbon reduction targets that lag behind current climate science or estimated impacts that underplay long-term impacts.

So, it is fair to assume that urbanisation *is* inevitable unless economics is changed. Current economic development strategies tend to create jobs and higher value-adding economic activities in cities, particularly those with global transport connections—whilst increasing inequality as migrants move into poorer living conditions in informal settlements that are often highly vulnerable to climate change. Globally, current patterns of development will lead population growth to remain focused on existing, including highly climate vulnerable, cities—either expanding principal economic hubs or expanding the population of commuters living in satellite cities and towns. There is a need to shift investment from increasing resource use, inequality and climate change to prioritising the creation of a climate-resilient, sustainable and fair future. This will require different economic tools, processes and decision making.

Automation and financialisation won't help address climate change

So how might widely mooted plans for increased automation, from driverless cars to industrialised agriculture, affect this? Most automation also requires continued or expanded infrastructure investment. Tim Jackson (2018) argues that digital and robot technologies may turn out to be in direct competition for investment with measures essential for the transition to a zero-carbon future. Jackson notes that investment in automation may remove the need for whole sections of the working population, while likely having a higher carbon and resource footprint (as more capital assets are required).

Paul Mason (2015) is optimistic that a technology-driven 'sharing economy' will empower workers to fight financial capitalists in new and powerful ways. However, Rana Foroohar (2016) is far less optimistic, noting how Apple itself can invest to sustain its own financial pre-eminence and how Uber's research into autonomous vehicles would benefit tech-giant owners at the expense of employees. Similarly, the expansion of renewable energy, including across Africa, is dominated by corporate interests, including oil and gas companies (CB Insights, 2018)—but this has tended to sit alongside existing fossil fuel investments, rather than transform overall development pathways (including urbanisation).

Another trend, exemplified by the level of debt up to and since the economic crash in 2008, is an increased financialisation of the global economy. Increasingly economic growth is in non-tangible assets, or asset price inflation (such as reflected in housing and real estate markets). Similarly, one climate insurance fund in the Caribbean recently returned over a third of the premiums paid as profit to the insurers—financialisation does not tend to put current, let alone future, generations' interests first (Jubilee Debt Campaign, 2018).

This increased financialisation of the economy is also reflected in more traditional sectors. For example, BP now makes over 20% of its profits from its financial trading division (Foroohar, 2016). This trend is increasing the importance of cities, and notably global financial centres in the world economy, and is likely to increase pressures to urbanise the economy. The IMF (2015) views infrastructure as an investment strategy to 'avoid stagnation'. But what if the scale of environmental pressures and climate change locally mean that an end to real growth is inevitable? Tim Jackson (2018) argues that the drive to seek economic growth (including through infrastructure-led investment that leads to further urbanisation and globalisation) is currently backfiring—widening inequality and locking-in carbon emissions instead.

As government keeps interest rates low there is increased investment and wealth accumulating to the richest, and increased investment in financial assets and speculation. This is increasing inequality and diverting finance from addressing the climate challenge. This in turn is suppressing spending in the productive economy, prolonging austerity, and so the vicious cycle carries on. Therefore, relying on automation or financialisation is not likely to address the way urbanisation impacts on and drives both climate change *and* inequality.

Facing up to climate change requires *different approaches at a national and global level*. It is insufficient to expect a clear climate-resilient pathway locally to be achieved while 'rules of the game' that externalise the climate and environment and don't factor in economic distribution are left to drive decision-making.

Taking a different way forward

Climate-led strategic planning

Attempting to reduce resource consumption and the related climate impacts whilst prioritising economic growth through further infrastructure investment and urbanisation is difficult, if not impossible. This is reducing the resilience and livelihoods of the poorest households, geographic areas and countries, and increasing inequality, social tensions and global instability (Sassen, 2014). Instead of city economies being increasingly *externally dependant* as they get bigger, there is a need to localise investment to city-regions, making them more *resilient*. So, what if instead of investing in infrastructure that primarily strengthens connections between urban areas, the focus was on strengthening links from cities to their immediate rural surroundings?

This section explores a different and longer-term plan for how societies can sustain drastic cuts to greenhouse gas emissions and thereby limit the increase in the scale and magnitude of future climate disasters.

Climate scientists say we need to avoid 1.5°C of post-industrial revolution warming to avoid dangerous climate change (IPCC, 2018). This requires a radical change to our approach to urbanisation, transport infrastructure investment and wider economic development. We have just 12 years to turn things around and reach zero fossil fuel emissions by between 2040 and 2055 at the latest. Some might say that this level of transformation is not possible. Instead we should start by asking whether living with dangerous, runaway climate change is any way acceptable or desirable.

Current climate trends will already lock in short-term impacts even if long-term drastic emissions reductions still take place. This is simply not factored into what we plan to build and where. The result could be catastrophic. Addressing small short-term climate impacts, such as incremental increases in sea defences will provide a false sense of security—while being completely inadequate in the face of more extreme disaster events and in the longer term. And if we make decisions about where and how we expect to live for future generations we should also consider the even longer-term consequences. For example, Stern (2007) concluded that:

> Recent estimates suggest that, even if emissions peak in the next decade or two and then fall sharply, the impact on global temperatures will still be very large [as] many greenhouse gases, including carbon dioxide, stay in the atmosphere for more than a century and the effects of climate change come through with a lag, temperature and sea level will continue to rise during the twenty-second century, even if we stabilise emissions soon.

The positive feedback effects within the climate system and the increased energy and unpredictability in a warming climate means that greater damage and climate shocks are increasingly likely with continuing global warming. Therefore, it is irresponsible to continue to increase carbon emissions now and gamble on greater reductions (including negative emissions) at some point in the future.

The aim should not *just* be to stay within a certain carbon budget, but to limit our exposure to climate risk by reducing carbon emissions to *as far as possible* and *as soon as possible*. This points to a planned approach.

Socially and environmentally sufficient planning

Kennedy and Corfee-Morlot (2012) propose that comprehensive national strategic plans are coupled with national climate change goals to prioritise investment differently. This requires *completely different* investment decisions—not just at the city level but crucially by countries and internationally (particularly as lower-lying and hotter countries become increasingly precarious and/or inhabitable).

The Green New Deal Group (2013) advocated investment to revitalise and transform local economies, including to rejuvenate declining rural and ex-industrial areas. Similarly, considering the future of the EU, Ulrike Guérot (2017) suggests moving away from a centralised economy and instead looking below nation states to decentralise politics as well as economies and energy systems. Such a transformed scale of energy and resource use could raise the well-being, equality, self-reliance and resilience of communities and economies (Scott Cato, 2012; Illich, 1973). This approach underpins the transition towns movement in the UK and elsewhere—which links the notion of energy descent with community resilience (Hopkins, 2008).

So, what if this approach focused on social and environmental objects together? The World Wildlife Fund (WWF) compared the different social and environmental outcomes of different countries, highlighting that it is possible to be sustainable environmentally whilst having good social outcomes (WWF, 2012). And this can happen quickly. Examples include Costa Rica, which generated 99% of its electricity from renewable sources in 2015 and Cuba, which five years after the oil embargo of 1997 had halved obesity and halved deaths due to diabetes whilst establishing 26,000 community gardens in Havana alone (Simms and Newell, 2018).

The pattern of development required for the poorest nations to be sustainable is entirely different from richer economies: the former must improve the livelihoods (and resilience) of the poorest whilst not increasing their (minimal) carbon emissions, whilst richer countries, urban areas and households must reduce their (still increasing) levels of consumption of environmental resources and carbon footprint whilst sustaining well-being. Dercon (2012) argued that it is best to reduce poverty first and consider climate and environmental issues later, but this risks locking-in climate emissions through infrastructure choices, as considered above.

So, facing up to climate change globally turns the challenge of making development (as a continuing process) more sustainable into one of finding different pathways towards the twin goals of environmental (resource and biodiversity) sufficiency and social (equality and quality-of-life) sustainability. The 'pathways' which Bangladesh and the UK must take (for example) are totally different, although the destination is shared.

If one were also to compare the social and environmental sustainability of human society at different *scales* (instead of in different places) we would see that it is easier for smaller communities and city-regions to become sustainable than mega-cities. Reducing the scale of the 'area of interest'—localisation—makes it easier to become more sustainable. This needs to combine with political, economic and ultimately cultural change—to define how we will be sustainable *enough* to avoid runaway climate change. Infrastructure investment would then be limited to that which returns a sufficient reduction in carbon emissions whilst reducing inequality and transforming the resilience of communities to withstand future disasters and climatic changes. This is a massive shift. It requires planned choices that deliver equality *and* well-being using *far fewer* resources *and* resilience.

One of the greatest challenges will be how to deliver such simultaneous climate adaptation and mitigation, whilst ensuring benefits are shared more equally. Research by the UK Climate Impacts Programme (Lonsdale, Pringle and Turner, 2015) concluded that we must shift from preparing for impacts predicted by old climate data—separate from climate mitigation plans that lack adequate ambition—to mitigating and adapting to climate change together. They call this *transformational adaptation*. In the long term a complete transformation of society, not just adaptation of existing activities, will be required.

This is reflected in the IPCC's special report into the implications of a 1.5°C rise in post-industrial global warming (2018). This considers the implications of climate change on poverty alleviation and development. It calls from a shift from such incremental approaches to 'transformational adaptation' through 'ambitious and well-integrated adaptation-mitigation-development pathways that deviate fundamentally from high-carbon, business-as-usual futures'.

The built environment must be designed to last *and* be more adaptive, both cutting emissions more quickly and reflecting a more pessimistic climate future. However, unless we carefully plan and constrain infrastructure investment (and hence urbanisation), focusing on increased adaptation alone could result in a huge up-front expenditure of carbon embodied in new 'built environment'—which will both make climate change worse and could alone tip us into catastrophic climate change.

Some rural and urban areas will become untenable due to sea level rise, increasing disasters and/or changes in food productivity or water availability due to climatic changes (see above). A planned retreat, initiating relocation from some are required. This will mean shifting building from areas that will flood or otherwise become uninhabitable in the longer term. It will entail migration from precarious livelihoods such as those on increasingly inundated land in south-west Bangladesh. And it will require changing what is normal, such as changing approaches to farming as weather patterns lead to escalating crop failures in sub-Saharan Africa. Transformation will require joining up issues that are often considered separately: equality, resilience (of communities and infrastructure) and sustainability (including climate mitigation). This requires a step-change, taking approaches that aim to benefit *all* of humanity.

This needs a plan: spatially, economically and strategically. So, instead of reinforcing existing settlement patterns, increasing city size and vulnerability—investment should strengthen ties to the local economies around urban areas. This will enable more people to stay living and working in smaller settlements instead of migrating to larger centres. This will help to placate the drivers of existing shortages of affordable homes that lead to informal settlements (particularly in the global South) and over-occupancy of housing (where land prices preclude affordable rents, such as in London, UK). However, significant magnitudes of migration will need to be anticipated and planned for. New homes and ways of life will need to be supported without the massive embodied carbon typically invested in current cities.

The countryside underpins the sustainability of cities as this is where most food, resources and energy are supplied from. If city dwellers' needs are met from areas immediately around cities, rather than across the world, transportation and packaging will be reduced. Alternative economic strategies shifting from investments that continue to enlarge and link global cities to recreating city-region economies (also referred to as bioregions) could transform rural livelihoods and improve resilience, reducing vulnerability to climate change. Plans to address this include that proposed in Rwanda, to shift the focus of growth away from the capital city, Kigali (GGGI, 2014).

Our infrastructure must be reimagined today such that it can be re-purposed and even moved as our society and economy are restructured and localised. Investment must be limited to *very* low and zero carbon technologies, energy efficient *enough* and designed-for-deconstruction (Addis, 2008). This means designing for a truly circular economy, not just in terms of what we reuse and recycle at home, but also in terms of our overall homes, infrastructure and wider built environment. We need to apply the principles of permaculture (permanence) so that we can sustain our ways of living with renewable resources alone—redeployed, again and again. For example, any new homes should be negative carbon (Rodionova, 2016) and if we do choose to build in flood vulnerable areas, such as the UK's Thames Gateway Development (why would we?), these homes must be able to be relocated within their lifetime.

A different future requires that we face up to two separate, yet interconnected, societal challenges, together. Currently we are not acting sufficiently to avoid runaway climate change, which risks skyrocketing inequality and societal breakdown. To cut climate change we must radically reduce resource inequality as over half of emissions are due to the lifestyles of the richest 10% of the global population (Anderson, 2018). This would require an alternative to the economic trends of automation and financialisation, and through these, continued globalisation of the economy. In place of often unplanned urbanisation we might then radically re-localise how we live.

Conclusion

Urbanisation is not inevitable. Facing up to climate reality requires us not to take urbanisation and globalisation as parts of our almost certain future, but see these as choices. This requires a radical shift in political, economic, and physical planning with different priorities and politics, tools and approaches. But more than that—it needs a vision of a future that is not increasingly reliant on infrastructure investment, automation or financial services.

As noted above, the IPCC (2018) calls this transformative adaptation:

> Climate resilient development pathways that transform societies and systems to limit global warming to 1.5°C and ensure equity and well-being for human populations and ecosystems in a 1.5°C warmer world …require ambitious and well-integrated adaptation-mitigation-development pathways that deviate fundamentally from high-carbon, business-as-usual futures.

Or to put it simply, we must take as our starting point the assumption that facing up to climate reality requires us to rethink, reimagine, change—everything.

Climate change is a messy problem: it cannot be tackled simply by the sum of seven billion people acting individually, according to enlightened self-interest, guided by the invisible hand of the market. Climate change challenges a rethinking of systems, from centralising and urbanising our economy to re-investing in our sense of place and definition of community. We need to restrict what we *physically* invest in to that which cuts our carbon emissions down to zero, whilst enhancing how we work together—and meanwhile not abandoning those facing climate disasters. Together there must be a massive transformation—a great flourishing and multiplicity of local solutions, economies and ways of living without oil that ultimately make us more reliant on each other instead.

So, facing up to the reality of climate change *together*, means that we commit that the sum of our collective actions turns out to be enough to stop runaway climate change. And therefore, *sufficient* to turn fear, insecurity and apathy into genuine hope. Foster (2017c) talks about finding deep hope in the space between our almost certain future with climate Armageddon locked-in and a future that is still paved with climate disasters, but not apocalyptic. If we fail to stop climate change's worst rampages, we will need to replace some of our physical infrastructure with reliance on our local and global environment and each other. That must ultimately be where we find ourselves, become more resilient, and build our hope.

Acknowledgements

Thanks to Ray Cunningham, Sarah Finch and John Foster for their edits and to Fiona for her support.

Chapter 5

DEALING WITH EXTREME WEATHER
Anne Chapman

Introduction

Storms, floods, droughts and heat waves—extremes in weather—have become more frequent in recent years. Between 1980 and 2016 the frequency of floods and other hydrological events quadrupled, while the number of droughts, heat waves and forest fires doubled (EASAC, 2018). The 2017 hurricane season in the Caribbean and Southern USA provided perhaps the most visually impressive examples of extreme weather and the damage it can cause. There were 10 hurricanes in a row, three of which, Harvey, Irma and Maria, together caused $282 billion worth of damage. Houston was flooded, several Caribbean islands devastated and Puerto Rico lost its electricity grid (2017 Atlantic hurricane Season, 2018). At about the same time, though receiving less media attention, over 40 million people were affected by flooding caused by monsoon rains in South Asia, which killed around 1200 people (2017 South Asian Floods, 2018).

The focus of this chapter is on two very different examples of extreme weather in Western Europe. It examines how these events unfolded, how people coped and reacted afterwards. The first, longer, account is of the impact of Storm Desmond on Lancaster in December 2015. Lancaster is a small city in North West England. The area covered by Lancaster City Council includes an almost contiguous urban area of Galgate, Lancaster, Morecambe, Heysham, and Carnforth, and a large rural hinterland with many villages. I live in Lancaster, so this is partly a first-hand account, drawing on my own experience, conversations with others in Lancaster and presentations given at a conference which I organised for Green House and the Green European

Foundation, in Lancaster in October 2017.[1] It also uses information from a report written by Professor Roger Kemp of the Engineering Department at Lancaster University, following a workshop he organised there in March 2016 (Kemp, 2016).

November and December 2015 saw a whole series of storms hit Northern England. Storm Desmond, the fourth named storm of the winter, resulted in the highest ever rainfall over a 24-hour period recorded in the UK: 341.4 mm falling on Honister Pass in Cumbria.[2] With that amount of rain falling on already saturated ground, roads turned into rivers, and in parts of Kendal people had water coming off the fell running through their houses. Most of the media attention was on places like Carlisle, Cockermouth and Keswick, which flooded for the third time in 10 years. In Carlisle several thousand homes were under water. A video was posted on Facebook of the water over-topping the flood wall in Keswick, built in 2012, after the 2009 flood, to withstand a 1:100 year flood. Boulders from landslips on the fells were washed into the village of Glenridding on Ullswater, causing the stream that runs through the village to overtop its banks, flooding nearby shops, houses and a hotel. Both roads into the village were flooded, cutting it off for three days. The stream burst its banks a second time four days later.[3]

Lancaster received less attention because the main impact of the floods was a power cut over much of the urban area in the district. The effects of a widespread power cut are less photogenic than flooded streets, and more difficult to report because of the difficulties in communication caused by the loss of power. For example, a BBC 5 Live reporter sent to Lancaster was apparently unable to broadcast because of the lack of mobile phone signal (Kemp, 2016, 8). This was the first major loss of power over an urban area in the UK since the power cuts of the 1970s and it revealed much about how dependent on electricity we have become.

The other example of extreme weather examined in this chapter is the heat wave in Paris in August 2003. The account here is mainly based on a presentation by Alice Le Roy at the October 2017 conference mentioned above (Le Roy, 2017). In 2003 Alice was a policy advisor to Paris City Council so had first-hand experience of the heat wave. Unlike a flood, a heat wave does not produce dramatic pictures, but can nonetheless have deadly consequences.

---

1 See https://www.greenhousethinktank.org/dealing-with-extreme-weather.html for information about this conference, including videos of the presentations.
2 See https://www.metoffice.gov.uk/public/weather/climate-extremes/#?tab=climateEx tremes.
3 See http://www.parishfloodgroup.org for more on what happened in Glenridding.

Both floods and heat waves reveal the importance of people helping each other, and looking out for each other.

Before examining these events I discuss how climate change is increasing the frequency and severity of extreme weather.

Climate change and extreme weather

It has long been known that global warming, caused by increasing atmospheric concentrations of carbon dioxide and other greenhouse gases, will increase the energy in our climate system, leading to stronger winds and heavier rain as well as higher temperatures. However, climate change will not merely mean a warmer world, but one with changed, and perhaps continually changing weather patterns. In the northern hemisphere the Arctic is warming faster than the tropics, reducing the differential between the two. This is causing the jet stream—the high altitude westerly winds that form the boundary between cold air to the north and warm to the south—to meander more than it did in the past. Rather than west to east winds forming a tight circle around the Arctic, the winds form a roller-coaster pattern, with more north-south and south-north winds. These winds block weather patterns, so that hot air, or rain, become trapped in the same place for weeks or months, rather than moving on (Sutherland, 2017). For example, in the summer of 2010 slow-moving meanders of the jet stream caused floods in Pakistan, which killed 2,000 people, and a heat wave in Russia, which killed 50,000 people (Marshall, 2010 and Mann *et al*, 2017). In the summer of 2018 they blocked the movement of high pressure which sat over western Europe causing a prolonged heat wave (Watts, 2018). The shifting jet stream can also bring cold air south and warm air north (Coumou *et al*, 2018). Thus in January 2017 cold air from Northern Russia moved south west to Southern Europe, with big snow falls in Greece and Italy.[4]

In the UK there have been several periods in recent years when a stationary jet stream has resulted in prolonged wet periods, one such period being in November/December 2015 over Northern England, a period which included Storm Desmond. Storm Desmond created an 'atmospheric river' which brought water from the Caribbean and dumped it on the British Isles.[5] Climate change is thought to have made Storm Desmond approximately 40% more likely (van Oldenborgh *et al*, 2015).

---

4 See https://en.wikipedia.org/wiki/January_2017_European_cold_wave.
5 See https://en.wikipedia.org/wiki/Storm_Desmond.

We need to remember that this is just the current situation. As the Arctic warms further, things are likely to shift again. Currently, ocean currents known as the Atlantic Meridional Overturning Circulation carry warm water from the Caribbean and Gulf of Mexico north, past Western Europe to the Arctic, where the water cools, sinks and flows back south. The northward current, the Gulf Stream, is what makes Western Europe much milder than similar latitudes in the Americas or Asia. However, as the Arctic becomes warmer and increased inflow of fresh water from melting glaciers makes the seas there less dense (salt water is denser than fresh water), the amount of water cooling and sinking is likely to decrease. With less returning water, the flow north would decline or even stop, which would have dramatic consequences for the climate of Western Europe. Two studies published in early 2018 suggest that this ocean circulation is weakening (Carrington, 2018 and also discussed in EASAC, 2018. See also Marshall 2018).

Lancaster and Storm Desmond

It rained in Lancaster on Saturday the 5th of December 2015. It had rained everyday but one for over a month—an unusual amount of rain even for the North West of England.[6] But this rain was exceptional. We went out twice to events at a local school, both times having to battle against the wind and navigate the swift flowing streams that had appeared at the sides of roads. A red flood alert was issued for the Northern reaches of the river Lune, which flows from the Howgill Fells into Lancaster. Parts of Lancaster near the river were issued with an amber alert, but many businesses in the riverside industrial area did not take this warning seriously: those that did went in on Saturday morning and moved critical equipment such as computers upstairs, saving themselves much trouble and cost afterwards.

Lancaster's Saturday went on as usual with people in town shopping, then as the evening wore on, enjoying a night out in the run-up to Christmas. The town centre is in a valley, with a steep hill to the east, the smaller hill of the castle to the west and the river to the north. Two road bridges, a foot bridge and railway bridge (with a footway alongside the railway) cross the river. Properties along the Quay next to the river are protected by a flood wall. This has gates in it which are locked open. At just before 5pm, as the river rose, a local councillor noticed that the gates had not been closed and water was nearly

6 Rainfall in November and December 2015, measured at the weather station at Lancaster University, was 556.2 mm, more than one and half times the previous record. There was only one day, 22nd November when it did not rain (Lambert, 2016).

up to the top of Quay wall. He phoned the Environment Agency emergency number who said that someone would come with the key, but could not say when. The councillor, not knowing when someone from the Environment Agency would turn up, went and spoke to the fire service who were pumping water out of nearby cellars, who got the police to come along with bolt cutters. The police the councillor and a fire fighter all waited at the gate, unsure how to close the gate if they cut the lock off. They did not know how long the Environment Agency would be and were unable to contact the people coming as the call centre staff would not give them their phone number. Finally, staff from the Environment Agency turned up at 5.45pm and shut the gates before the water came in from the river.

However, the water was coming round the back, up through drains, flooding the Quay from the land side. There was so much water pouring down through town unable to get into the river that it built up on the town side of the flood wall. It is also thought that drainage from an old, now buried mill race that flows beneath the bottom of the town, contributed to the flooding. With the bottom of town flooded it was impossible for people to get to their homes north of the river. The police were busy trying to persuade drunken people from venturing into the floodwater and buses were laid on from higher up the town to take people home to Morecambe (4 miles away), and other places north of the river via the motorway, a journey of over 20 miles. Others ended up staying in Lancaster sitting in pubs with blankets to keep warm. Then, at shortly after ten thirty, the lights went out.

Lancaster's main substation is on the flat land next to the river which stretches north-east from the town centre. It is on the site of the former coal-fired power station, built in 1915. Its location allowed it to use river water for cooling and to be supplied with coal by the railway from Yorkshire which ran between it and the river. The power station closed in 1975, but nobody thought that the substation which remained should be moved out of the flood plain. The last major flood had been in 1907. On that Saturday Electricity North West, the private company which owns the low voltage electricity grid in North West England, had been working all day pumping water to try to keep the sub-station going. At 10.39 they gave up and turned the substation off, cutting off supplies to most of the urban area of Lancaster, Morecambe and Carnforth.

Sunday the 6th of December dawned bright and clear. In our home high on the hill on the east side of Lancaster we were unaware of the flooding. We did have a power cut and as my son and I walked across town to get to the Quaker Meeting, we realised that everywhere was without power. We met someone we know desperately trying to buy matches—almost all the shops

97

were shut. At the Meeting House there was a short half hour meeting in the cold, sun-lit room, sitting with our coats on, and no tea or coffee afterwards. We then walked down to the river to see what was going on. It felt like half of Lancaster was down there, enjoying the sunshine and staring at the raging torrent that the river had become. Someone said it was low tide, when the river is normally twenty foot or so below the wall we were on. Now it was just below it. The water had mostly receded from the streets but the basement garages of flats by the river were full of water with submerged cars. Sainsbury's supermarket is right next to the river and there was a big queue of people outside: the store had been flooded and they were giving food away. With their huge resources Sainsbury's were able to clear the store, clean up and re-stock within a day or so. Other local businesses were not so lucky, remaining closed for months.

Police were at the entrances to the footbridge and two road bridges, stopping anyone crossing. They told us that some shipping containers had come down the river during the night and hit the bridges, so they were closed awaiting inspection to ensure that they were safe. That is also where we learned that the reason for the power cut was the flooding of the substation, and that it would be three days before it was operational again.

Having no electricity for a day or so makes you realise how dependent on it we have become. We were relatively lucky: we have a wood burning stove and a gas cooker, so were warm and could cook. People with only electricity were less fortunate. They had no heating or cooking facilities and for those in flats with pumped water supply, no water. Those living at home with care needs were particularly vulnerable, with stair lifts, home dialysis, oxygen therapy and alarm systems all dependent on electricity, and on the mobile phone network for communication with carers.

I know someone who lives in the upper floors of a block of flats and whose husband requires oxygen. His oxygen machine requires electricity. When the power went off they used their emergency cylinders, but by the Sunday morning they were close to running out. She was advised to go somewhere else for a few days if she could. With great difficulty, and help from a friend, she managed to get her husband and the oxygen machine down to the car (the lift was not working of course) and drive to a hotel about 10 miles away. She was worried she did not have enough fuel in the car (the local petrol stations were closed), but fortunately made it. She had no contact from social services who were running an emergency centre for vulnerable people, though this was north of the river, so difficult to reach from where she lived, south of the river.

For us the main issue was communication: there was no internet, mobile phone signal or landline phone (because we did not have a fixed-line phone which plugs in directly to the phone socket—we have now bought one!). We only found out what was going on by going out and asking people we met on the street. By the Monday night though we realised that the local commercial radio station, Bay Radio, was broadcasting on their FM signal with information about what was happening (there was no digital radio signal). They have a studio on the Quay which was flooded, but they managed to move things upstairs and got someone they knew in Lancaster to set them up with a generator with which they were able to broadcast. They sent a reporter to somewhere unaffected by the power cut who then phoned them on their landline with information gleaned from the internet and local people phoned in with information. We listened to it on our battery-powered radio, to find out whether the school was going to be closed for another day, while playing Monopoly by torchlight.

In the 1970s, when Britain last had widespread power cuts in urban areas, doors were opened by hand and locked with a metal key or bolt. Now, electrically-controlled doors abound. Some people may have had all sorts of useful stuff, like camping stoves, in their garage, but could not get into them because they had a powered garage door with no side access. I heard some months later that a group of men went to the ambulance station with pick axes to break down the door there so the ambulance could get out. A couple of months after the flood I spoke to a couple who had tried to visit an elderly person they knew who lived in a flat near the town centre, in the part that was flooded, to check that she was OK. They lived north of the river so on the Sunday they walked over the railway bridge on the footway. However when they got to her flats there was no way to get in or contact her: the door was locked and the intercom not working.

We have also linked up all sorts of systems to ensure safety in a way that relies on electricity, but which cannot be over-ridden in times of need. For example, one of the care homes had a gas hob, but this could not be used if the extractor fan (powered by electricity) was not working. So instead the chef cooked on a BBQ in the garden and some relatives of a resident bought in a camper van with a gas cooking stove. At another care home the lock on the back door was linked to the fire alarm. With no electricity it could not be locked, so someone had to sit next to the door all night to ensure no intruders came in (BBC, 2018).

At Lancaster University there were rules against people sleeping in rooms without fire alarms and emergency lighting. These were ignored on the Saturday night. It was a week before the end of the university term so on the Sunday, when they thought that the power was going to be off until Tuesday evening, it was decided to end the term early and send students home. However, on that

Sunday the long distance trains on the West Coast mainline were not running north of Preston and after dusk no trains stopped at Lancaster station because there were no lights on the platforms (fortunately the power for the trains was from outside the affected area). The University put students who wanted to leave on buses and took them to Preston station, where they were left to fend for themselves. This transferred the responsibility from the University to the students, but would they have been safer staying in rooms without fire alarms?

Students living in town also had a difficult time. The fact that many did not have stocks of food, but just bought what they wanted each day was revealed by the problems they had when all the shops were closed. They also had to learn how to use a public telephone: there were reportedly lines of students outside phone boxes. The decision to close the University a week early reached students in town by word of mouth (the normal method of communication, email, being out of action). Inevitably there was corruption of the message: several hundred overseas students arrived at Lancaster police station expecting to be provided with transport.

The hospital was better prepared. It had generators and had recently had an emergency practice. Its cafeteria did a roaring trade on the Sunday, being one of the few places where food could be bought. There were stories of students going in with extension leads to charge their phones and laptops.

For us electricity came back on in the early hours on Monday morning. This was provided by a truck-mounted generator parked outside the substation round the corner from our house. Seventy-five generators were brought to Lancaster from all over the country. Electricity North West was not guaranteeing the supply so most local schools made the decision to close for three days. There were further power cuts, with different areas losing power at different times. For us power went off again on Monday afternoon, but was on by Tuesday.

One thing that became apparent at the workshop convened by Roger Kemp in March (Kemp, 2016), is the diffusion of responsibility for some of our critical systems. Privatised public services and utilities and the various bits of the public sector, each look after their own bit of the system, but no one seems to have overall responsibility. There was a 'Gold Command' in Lancaster, co-ordinated by the police, but this only communicated with Electricity North West and the emergency services (fire, police and the hospital). People like head teachers and University vice chancellors were very much on their own, having to make decisions with no guidance. The Vice-Chancellor of the university, for example, was not contacted by anyone and was reduced to phoning people who lived outside of Lancaster to get them to look things up on the internet to find out what was going on.

There is also the issue of the geographic spread of responsibilities when communication is difficult, and the lack of someone on the ground empowered to take decisions to do things that are contrary to normal procedures. The alarm system for a local primary school, for example, was monitored by someone in Belfast who could not communicate with the head teacher in Lancaster. The manager of Lancaster train station is an area manager based in Carlisle. Booths supermarket in south Lancaster was one of the few shops open because they had a back-up generator. They consequently did a roaring trade on the Sunday, but despite the demand closed at 4pm to comply with Sunday trading rules, there being no one local able to make the decision to stay open.

The power cut could have had worse consequences: water, sewage and gas systems all continued working as equipment requiring power was either outside of the affected area or had back-up supplies; the main day of the power cut was a Sunday, when most people were not at work and schools were closed; after the first day power was supplied by 75 mobile generators, brought in from all over the UK, which would not have been possible if other areas had also been affected by power cuts.

Unlike Carlisle, where significant residential areas were flooded by Storm Desmond, in Lancaster the number of residential properties affected was relatively small and it was primarily businesses that were flooded. The topography of the town meant that the extent of flooding was limited. However, everyone was affected by the power cut. This illustrates that while concerns over flooding tend to concentrate on the experience of those flooded, which is often devastating, the number of people involved is generally relatively small. Secondary impacts, such as Lancaster's power cut, tend to affect far more people. In Cumbria, for example, Storm Desmond washed away part of the main north-south road in the Lake District, linking Grasmere with Keswick. This turned a 20 minute car journey into a 1.5 hour trip. Some school children who went to school in Keswick but lived in Grasmere reportedly had to stay with friends in Keswick during the week (Szönyi, May and Lamb, 2016, 23) and the cost to the local economy was estimated at half a million pounds a day (according to a report in the *Westmorland Gazette* on 10 May 2016). The road was not reopened until May 2016.

A year on from the Storm Desmond 19 of the 212 businesses affected in Lancaster were still not in operation, 22 were only partially operating and six had permanently closed or moved out of the district (Rouncivell, 2016). Two years on the *Lancaster Guardian* (on 7 December 2017) reported that the substation had been raised 10ft (3m) off the ground and there were plans for a flood wall along the Lune upstream of Lancaster centre to protect the adjacent industrial area. Funding for the flood wall had yet to be found as commercial areas are given a lower priority than residential ones in the funding formula used by the Environment Agency to allocate funding to flood-prevention schemes. However, little had been done to improve the surface water drainage system in Lancaster, though there had been investigations into the old mill race. Torrential rain on the evening of the 19th of July 2017, for example, caused flash flooding that affected many of the businesses in the lower part of the town centre that had been flooded by Storm Desmond.[7] However, at least one of these was open soon after because when reinstating their food takeaway after Storm Desmond they had done things like moving the electric sockets up the walls and putting in a flood-resilient floor. There was also flooding in the lower part of the town, and in villages to the south and east of Lancaster, following record rainfall associated with Storm Brian on the 22nd of November 2017. These events have raised concerns about the impact on flooding of new and planned housing developments (Lakin, 2017).

These more recent events have been caused by local heavy rain, whereas the flooding caused by Storm Desmond was an outcome of a month of high rainfall over the whole catchment of the Lune, increasing the flow in the river so that it burst its banks. Many think that the management of the uplands needs to change, to increase their capacity to retain water, reduce erosion of peat and reduce flooding downstream. Large scale tree planting has taken place in the Howgills, co-ordinated by the Woodland Trust, and at least one local farmer near Lancaster is working to increase the water retention-capacity of his moorland, and encourage others to do the same (Leeson and Everett, 2017).

Storm Desmond sparked an interest in emergency planning in Lancaster District. Prior to December 2015 the only community with an emergency plan was Sunderland Point, an isolated hamlet at the mouth of the Lune Estuary which is vulnerable to coastal flooding. Now several villages have plans. The focus has inevitably been on what would happen if they lost electricity, with local village halls applying for funds for generators so they can function as an emergency centre in the event of a power cut. The emergency planning officer at Lancaster has been working with these groups, helping them to write an emergency plan and train wardens to check on vulnerable people in their homes. His aim has been to make local communities more resilient and self-sufficient. In villages parish councils have often taken the lead. In urban areas it has been more difficult and the focus has been on setting up existing community buildings to be emergency centres, rather than getting people to

---

7 See https://www.lancasterguardian.co.uk/news/1-8m-shortfall-as-fears-mount-over-lancaster-flooding-1-8680248.

go out and check on people in the community. In Lancaster at least two community centres and a local church are now set up to be emergency centres, but in some parts of the district, particularly in Morecambe and Heysham, there is a lack of suitable community buildings (Bartlett, 2017).

Following the flooding in 2017 Lancaster Area Search and Rescue has been set up, as a charity, by local people. Their aim is to have a trained flood rescue team, with a boat and 4x4 vehicles in Lancaster, to assist the emergency services in the event of another flood.[8]

Paris Heatwave

Monday August the 4th 2003 was a beautiful sunny day in Paris. Children played in the fountains in the Trocadéro gardens near the Eiffel Tower and people enjoyed the pop-up beach on the banks of the Seine. It was hot, 30°C, but this was not taken as a cause for alarm. News reports talked about the good weather, competition for parasols and the importance of using sunscreen.

The following week was the hottest week in France since at least the sixteenth century. The heat was caused by strong southern winds and hot continental air that stayed over Western Europe. In Paris temperatures remained at over 30°C and did not fall significantly at night (lower night time temperatures are important to enable us to sleep). Northern France is unused to such prolonged high summer temperatures and there were no plans in place to deal with the situation, high temperatures not being considered a major hazard.

The first indication that something was wrong came on the 11th of August, from a doctor in charge of an emergency room at a Paris hospital. He was filmed by several television crews inside the hospital, with patients on trolleys in a corridor in the background. He described the situation as absolutely dreadful, with alarming numbers of people admitted to hospital with heat stroke, dehydration, confusion and similar conditions. Meanwhile, the deputy Mayor in charge of the Environment, whose portfolio included the cemeteries, was aware that something was wrong because the funeral services had told him that they could not cope with the numbers of bodies they needed to bury. He called for the resignation of the health minister, but many thought that he was overreacting. The government denied that there was a problem.

However, on the 8th of August the Health Service had brought in an extreme heat plan, calling nurses into work and delaying non-urgent admission. It was later realised that even by the 4th of August there were 300 more

8 See https://www.lasar.org.uk.

people dying daily than would normally be the case. By the 11th of August, after a week of high temperatures, that had risen to 2000 excess deaths per day. It is thought that on the night of the 11th of August, 'black Monday', 3000 people died in one night, more than on any night during the second World War when Paris was bombed. The numbers of bodies were such that a refrigerated warehouse at the edge of the city was used as a temporary morgue (BBC, 2007). On average the death rate was 60% higher than usual and throughout the whole of France 15,000 people died. Overwhelmingly those who died were elderly people living alone. It was realised afterwards that being alone was the key factor: those in nursing homes, who were generally frailer than those living in their own homes, were less likely to have died because they had people caring for them, who could ensure that they kept cool and hydrated. The fact that so many people in Paris go away on holiday in August had not helped. The impression was of younger people going away on holiday, leaving their elderly at home on their own to die in the heat.

In France there was much argument when the extent of the death toll became apparent. The President, Jacques Chirac, returned from a three-week holiday in Canada at the end of August. He called for more solidarity among French people. People felt that he was criticising the population and not accepting his own responsibility. The government blamed the health service, the health service blamed government complacency. The Surgeon General resigned and the health minister was dropped from the cabinet six months later.

In 2004 France brought in a heat wave plan. This has four levels of alert: green, yellow, orange and red. The plan involves: identifying at-risk persons, informing the public when high temperatures are expected and what measures they should take to keep cool, and creating cooled rooms. In Paris at-risk persons are registered on the 'Chalex' file (for *chaleur extrême*) and during high temperatures are contacted by phone every 48 hours to find out how they are and remind them of the measures they should take to combat the heat. After the 2003 heat wave it was realised that city residents are more at risk from heat waves than rural ones because of the urban heat island effect.[9] Temperatures in urban areas are higher than in surrounding rural ones because concrete, brick and tarmac provide dark surfaces that absorb rather than reflect the heat and less vegetation means less cooling from evapotranspiration. The difference between urban and rural areas tends to be greater at night than during the day: in the city during a heat wave the night-time drop in temperature which is so important for sleep and for health is less than in a rural area. In Paris after 2003, among the short-term measures taken to offset the urban heat

---

9 See https://en.wikipedia.org/wiki/Urban_heat_island.

island effect was the decision to keep the city's parks open during the night, especially for those who were trapped in overheated apartments.

A connection with climate change was realised several years after 2003. It is now thought that by 2070 France will have a heat wave comparable to that of 2003 every other year. In 2007 a comprehensive Climate Change Strategy (*Plan Climat parisien*) was implemented. It included low-tech solutions to the urban island effect, such as equipping public buildings with window screens and planting thousands of trees along the city's streets and avenues.

Eric Klinenberg, in his study of the 1995 heat wave that killed 739 people in Chicago, has called heat waves 'silent and invisible killers of silenced and invisible people' (Klinenberg, 2015, 17). He was interested in why neighbourhoods that, according to the statistics on poverty and crime, were very similar, had very different death rates. On visiting those neighbourhoods he realised that what distinguished those where few people died from those with high death rates was 'social infrastructure', described by Klinenberg as 'active commercial corridors, a variety of public spaces, local institutions, decent sidewalks, and community organizations' (Klinenberg, 2004). In those places people went out and mixed with each other. Neighbours knew who was vulnerable to the heat and kept an eye on them. In contrast, in neighbourhoods that had lost their social infrastructure, characterised by boarded up buildings, vacant lots and broken, uneven properties, the elderly were afraid to go out. They stayed in their homes which became like brick ovens, and were likely to die of the heat.

Klinenberg contrasts the attention paid to 'spectacular and camera-ready disasters such as hurricanes, earthquakes, tornadoes and floods' with that paid to heat waves, which in America kill more people than all other extreme weather events combined.

> Heat waves receive little public attention not only because they fail to generate the massive property damage and fantastic images produced by other weather-related disasters, but also because their victims are primarily social outcasts—the elderly, the poor, and the isolated—from whom we customarily turn away.
>
> — Klinenberg (2015, 17)

How to cope with extreme weather

What can we learn from these two contrasting incidents of extreme weather about how we can cope in the future, as weather patterns change? Adaptation is the term normally used for this and the focus tends to be on physical infra-structure: on building flood defences, making buildings more resilient to flooding or easier to keep cool in the heat. Adaptation suggests a gradual pro-cess of change to better accommodate to an environment. Coping is perhaps a better term for how we might attempt to deal with the shifting extremes of weather we are likely to face.

Planning and preparation are certainly important. Having a plan for what to do in particular circumstance helps us to feel in control and to act. Being well prepared and having a plan can have a real impact in enabling people to be rescued from flood water, have enough to eat when cut off, or be kept cool in the heat. However, things do not always go as planned, and it is also impor-tant that people take initiatives in the situation they find themselves in, even where this goes against normal procedures. Thus in Lancaster the chef who cooked on a BBQ outside, and the people who brought their camper van to the care home to provide hot drinks were acting on their own initiative to deal with the situation, as were the men who went to the ambulance station with a crow bar to open the electric doors, the pub landlords who provided blankets and food to stranded people, staff at the local radio station who managed to keep broadcasting, and Professor Roger Kemp who afterwards organised a seminar to find out what had happened. However, the geographic dispersal of responsibility at a time when communication was difficult sometimes meant that those on the ground could not take initiatives, such as the supermarket that shut at 4pm to comply with Sunday trading laws, rather than staying open to meet the demand.

The loss of power in Lancaster shows that we need to ensure that we can cope without electricity. Doors should have manual overrides, so they can be opened, closed, locked and unlocked in a power cut. We need to be able to override safety systems such as the one not allowing the gas cooker in the care home to be used without the extractor fan. We need to hold on to corded, plug-in phones, and have things like emergency plans and key phone numbers on pieces of paper and in address books, not just online or stored on electronic devices.

However, the most important thing we need is a good community: a com-munity where people help each other in times of need. Eric Klinenberg has argued that new infrastructure to protect ourselves against climate change needs to support the creation of vibrant neighbourhoods, not undermine them (Klinenberg, 2016). He quotes as an example the proposals for sea defences for Lower Manhattan.[10] Rather than a massive sea wall, which would have

---

10 See http://www.rebuildbydesign.org/our-work/all-proposals/winning-projects/big-u.

cut people off from the coastline, the plans are for parklands and recreation areas at the water's edge, which will also protect the city from storm surges. Klinenberg argues that the social infrastructure function is just as important as keeping the sea at bay. Building communities should be an aim of all developments and planning more widely: it is an essential component of how we are going to be able to survive climate change.

Acknowledgements

I would like to thank all those who participated in the conference, *Dealing with Extreme Weather*, which took place in Lancaster on the 28th of October 2017. In particular: Alice Le Roy, whose presentation about the Paris Heat Wave I have used in writing the account in this chapter; Roger Kemp, who talked about the power cut, and Caroline Jackson who spoke about the impact of Storm Desmond on Lancaster. I would also like to thank the Green European Foundation, on whose behalf I organised the conference.

Chapter 6

GEOENGINEERING AS A RESPONSE TO THE CLIMATE CRISIS: RIGHT ROAD OR DISASTROUS DIVERSION?

Helena Paul and Rupert Read

Man has lost the capacity to foresee and to forestall, he will end by destroying the world.

— Albert Schweitzer[1]

Introduction

The 2015 Paris Agreement on climate was hailed as a successful breakthrough in the process of addressing anthropogenic climate change. However, the truth is that the agreement is hollow, anthropogenic climate change is accelerating dangerously and little real action is being taken, action of the kind and at the scale that would actually measure up to the threat. Instead, there is a desperate search for any kind of 'solution' that avoids having to reduce emissions and collectively tackling our deeply fossil-energy-dependent model of 'development'. In fact, the Paris agreement contains major loopholes and a central one of these is its tacit reliance on geoengineering. This involves basically two approaches: proposals for technologies to (a) reduce incoming solar radiation, that is: reduce the heating effect of the sun and (b) remove greenhouse gases from the atmosphere. However, the agreement fails to ask what are the risks involved in such approaches, and above all, who decides—and who has the mandate—to take those risks, which involve the whole planet

---

1 Rachel Carson dedicated *Silent Spring* to Albert Schweitzer with this quote from him. It is said to come from a letter he wrote to a beekeeper suffering losses due to pesticides.

and all of humanity, present and future. Can seeking to engineer the climate possibly ever be consistent with precaution?

Our key claim in this chapter will be that to gamble on geoengineering is precisely to *avoid* facing up to climate reality.

Where are we now after Paris (and Katowice)?[2]

Let us start by asking bluntly what it *means* to face climate reality, in relation to our topic in this chapter. What if mainstream assumptions around action on climate change actually embody tacit denial of its reality?

Consider the hard realities around the 'successful' Paris climate change accord:

- The Paris Agreement's targets are inadequate safely to address what the Agreement itself considers 'dangerous climate change', 1.5–2 degrees of overheat compared with preindustrial levels, because they give us at best a 66% chance of reaching those targets. (Imagine being asked to board a plane with a 66% chance of safely reaching its destination.)
- The Agreement has an inadequate definition of 'dangerous' climate change: for, as increasing extremes of weather underscore, it is clear that the climate change unleashed by human action is already dangerous, and we're not yet even at 1.5 degrees.
- Pledges from countries are entirely inadequate to reach the Paris targets— they head us instead toward probably 2.7–3.4 degrees of overheat (New Scientist staff and Press Association, 2016).
- There are in most cases no clear plans for how countries will reach those pledges, and virtually no plan is legally binding (Britain's Climate Change Act remains a rare exception; and its enforceability is arguable. Meanwhile, Britain is 'on schedule' to miss its climate targets without urgent action (Carrington, 2018)).
- Virtually all countries have economic, industrial, agricultural and transportation policies, plans and practices that directly *contradict* their stated aspirations to tackle man-made climate change.

---

2 See the article 'The Paris Climate Accords are starting to look like fantasy' (Wallace-Wells, 2018a). Moreover, 'progress' at Katowice seems to have gone into reverse.

- As implied above, and, crucially for this chapter, Paris's achievability rests additionally on 'Negative Emissions Technologies' (NETs), aka geoengineering (Anderson, 2016).[3]

We consider these points in detail below, especially in relation to the Precautionary Principle, but we start by making the following elementary observation: as yet, there is very little reason indeed to believe that (even if these NETs were acceptable philosophically or ethically) they will actually *work* even on their own terms (Wallace-Wells, 2018b; see also Proctor et al., 2018).

There is instead increasing reason to believe that they will not be economically or technologically viable, nor constitute worthwhile returns on energy invested (Radford, 2017). We are gambling the future of the human race on non-existent technologies which are quite likely not to work even on a best-case scenario. One might even, only slightly tongue in cheek, make the claim that 'Non-Existent Technologies' is a more accurate rendition of the 'NET' acronym...

- There is no enforcement mechanism for Paris.
- And, again crucially: in order to reach its inadequate and unenforceable targets, the Paris Agreement (Article 6) proposes the voluntary use of Internationally Transferred Mitigation Outcomes (ITMOs) that are meant to represent improved offsetting mechanisms (Rabinowitz, 2017). Article 6 point 4 of Paris mentions the establishment of a 'mechanism' 'to contribute to the mitigation of greenhouse gas emissions and support sustainable development' to be established by the Conference of the Parties (COP) to the United Nations Framework Convention on Climate Change (UNFCCC). Once again, such a mechanism as this could serve to distract attention from the fundamental need to *reduce* emissions drastically and in a sustained manner.[4]

The conclusion one has to draw from all this is unattractive but unavoidable: the Paris targets will not be achieved. Within a generation or less, we will very probably be facing an *exponential* increase in climate disasters—with inexorably rising tides, and global temperatures heading up toward 3 or 4 degrees of global overheat, a level incompatible with civilisation or human

---

3 Some will claim that NETs are not geoengineering. We consider the use of the term NETs little more than a marketing rebrand to escape the justifiably negative connotations of 'geoengineering'.
4 Thus this clause might be dubbed the 'Hopeful Houdini' clause.

'development' as we know it (Spratt, 2010). Furthermore, a key reason why some scientists and Paris have somewhat arbitrarily picked 1.5–2 degrees as the maximum 'safe' limit of temperature increase is simply that, above this, we are *likely* to face escalating feedbacks, possibly leading to runaway climate change. These feedbacks are many, including albedo loss (less sunlight reflected back toward space due to dust and snow melt), the disastrous consequences of the die-off of the Amazon rainforest, and the accelerating release of the ultra-potent GHG methane into the atmosphere.

If Paris is as good as it gets, then the going is going to get very bad indeed. It is therefore important that the Intergovernmental Panel on Climate Change (IPCC) report published in October 2018 (Allen et al. 2018), makes it clear that there is a big difference between the impacts of 1.5 and 2 degrees and that we must urgently commit to 1.5 degrees and no higher. Even 1.5 means serious disruption of ecosystems and major challenges to which biodiversity generally and humanity in particular will have to adapt. And, especially given the recent failure of international will at the Katowice COP, to achieve 1.5 degrees is virtually inconceivable.

Why do we say that Katowice was a failure? It was the chance for the world to *embrace* the 1.5 degree target, and its eye-watering consequences— but this did not occur. To the contrary, that target was in effect rejected, as a result of the concerted action of the USA, Russia, Saudi Arabia and Kuwait.[5] There can be little starker proof that the world is not even going to aim at 1.5 degrees.

And this, of course, is a key reason why Green House is seeking to focus our collective attention on facing up to climate reality—on how bad that reality now is and is set to become.

The longer view—past and future

Let us put this into a longer context. The discovery and exploitation of fossil energy could turn out to be the greatest temptation in human history. The 'Industrial Revolution' that began in the 18th century marked the beginning of measurable human-induced climate change, perhaps our first really perceptible long-term mark on the planet. In its absence we just might have succeeded in living in balance with this planet's myriad other inhabitants and extraordinary, intricate and subtle systems. Instead we are moving swiftly towards

---

5 See https://www.middleeasteye.net/news/saudi-arabia-and-kuwait-bid-block-un-endor sement-global-warming-report-1251204896.

an almost-unprecedented extinction of species and we humans are seriously damaging and degrading most of the ecosystems on which our lives depend.

Shockingly, it is clear that, while we could still potentially avoid or at least mitigate some of the worst problems we face if we reduced emissions of GHGs now, in 2019, (a process which we should really have begun back in 1990), *we show no real signs of actually deciding to do so.* Instead we tend to turn to 'solutions' such as geoengineering, which (as we will sketch below) would likely add to our problems—as well as for the most part being heavily dependent on the large-scale use of still more fossil energy to develop and deploy.

Our line of thought in this chapter therefore issues in a radical suggestion: it's time to wake up and embrace a new *form* of development, most of the elements of which are already in operation somewhere on the planet, either in cultural memory or indigenous and local practice. This radically revisioned idea of development involves reducing our consumption of resources and embracing the idea that human development is not the same as economic growth, nor is it dependent on high energy consumption and fossil fuel dependent technologies. Such a shift could also help to address the injustice and inequality built into the current model of development that is destroying our companion species and all our habitats, and that could soon destroy us.

However, the likelihood of this sensible, truly radical path being taken in good time is slim indeed if we persist in considering geoengineering as a potential escape from our current plight. For this is the proposed *substitute*, in effect, for the kind of action that is actually needed.

What *is* geoengineering?

Geoengineering, according to the Convention on Biological Diversity which has discussed the topic at length, is:

> A deliberate intervention in the planetary environment of a nature and scale intended to counteract anthropogenic climate change and its impacts.

Fundamentally, geoengineering involves two basic ideas: 1. diminishing the amount of sunlight reaching the earth (Solar Radiation Management or SRM) by blocking the sun's rays or reflecting them back into space; or 2. removing greenhouse gases from the atmosphere by capturing and burying them in the sea, in the earth, in fast growing trees, in old coal mines and oil wells, etc.—and hoping they stay there—(Greenhouse Gas Removal or GGR—also called Carbon Dioxide Removal or CDR).

Proponents claim that, even if we were to stop emissions now, there is a huge amount of CO_2 already circulating in earth systems that will continue to push up temperatures for some years. Thus we need to block sunlight or remove greenhouse gases as well as cutting emissions of them.

We agree that there need to be some efforts at carbon-removal. Centrally, we need to restore wild biodiverse carbon-rich ecosystems.[6] Doing so would involve us in *reducing* our impact on ecosystems, and placing them in a position potentially to flourish, whatever we do—however well or badly our species fares—in future. But the kinds of interventions involved in geoengineering all involve us in *increasing* our would-be domination of the planet via a fantasised control of planetary systems.

This *is the respect in which geoengineering is fundamentally not precautionary.* (We will expand on this point further in the next section.)

Let us now examine the methods collected under the heading of 'geoengineering' in a little more detail. SRM techniques include measures to increase surface albedo—basically the whiter a surface is, the more sunlight it will reflect back into space. Ideas range from painting most roofs white (a harmless and probably helpful suggestion, but too small-scale to constitute engineering the climate), to spreading white plastic over deserts, to developing genetically modified or gene edited crops engineered to be greyer leaved and hairy, or even cutting down boreal forests so the snow can better reflect the sun's rays out into space.

There are also proposals to reflect sunlight back into space before it hits the planet. These include increasing the reflectivity of clouds by continuously spraying salt water into them as they form; inserting particles into the stratosphere on a continuous basis to mimic the effect of volcanic eruptions; or sending gigantic mirrors or sunshades into space to shade or to reflect sunlight away from the planet. Particle insertion and salt water spraying would have to be maintained on a continuous basis, as their cessation, especially suddenly, could lead to even worse impacts than not doing them at all; we'll return to this point below. Injecting particles into the stratosphere might have disastrous side effects, for example halting or disrupting the monsoon cycle, with impacts on millions of people, their food security and biodiversity in general.

GGR techniques involve capturing and sequestering greenhouse gases, using different techniques to take CO_2 directly from the atmosphere, for example through reforestation and afforestation, the former being the

---

6 See the 'Restoration' section of Read & Rughani (2017).

restoration of forests that have been lost, while afforestation is the mass planting of trees in areas where they have not grown in the recent past. Other proposals include capturing CO_2 in specially constructed 'trees', known as direct air capture. Captured CO_2 must then be stored where it is unlikely to leak out, for example in the strata of near-exhausted oil fields. This is familiar to oil companies as it is already used for squeezing the last oil from such reserves. This GGR approach is known as carbon capture and storage. It has been combined with the idea of growing vast plantations to become a proposal for bioenergy with carbon capture and storage or BECCS. This would involve growing huge numbers of trees and other crops that absorb CO_2 as they grow. These would then be cut, burned in power stations and the resulting CO_2 captured and buried in old oil wells and other geological strata, as well as being used in certain industrial applications or in greenhouses to accelerate plant growth. Indeed proposals for using the CO_2 captured are becoming an industry in themselves, called Carbon Capture and Utilisation and Storage (CCUS) (Bio-Based News, 2018). The use of BECCS was assumed to be essential by the IPCC's A5 report and could be included under the 'mechanism' in the Paris Agreement. However, the recent Special Report on Global Warming of 1.5°C from the IPCC (Allen et al, 2018) seems to reflect revised IPCC opinion. It is critical of BECCS, noting that it would require between 25–46% of arable and permanent crop land on the planet, while BECCS plus afforestation might require all of such land, leading to untenable trade-offs for example with food production. It would also require a large expenditure of energy to put in place, together with major inputs of fertiliser.

Another proposal for GGR geoengineering is enhanced weathering. This entails mining, crushing and spreading of silicate minerals, to be broken down into carbonates by wind and rain, pulling carbon dioxide from the atmosphere and storing it in the soil and eventually the oceans (Beerling et al, 2018). However, there are questions about how long it will remain stored and about the efficacy of its potential soil and ocean co-benefits. For the technique to make a significant contribution to global mitigation efforts, major—carbon-heavy—infrastructural development and energy use would be required to mine, crush and transport the rocks.[7]

Other GGR approaches also involve using the soil as a carbon sink; for example, adding huge amounts of 'biochar' (industrially-produced charcoal) to

---

7 There are further questions about the release of toxic substances with potential human health impacts, particularly if dunite, also known as olivinite, is used (Strefler et al., 2018). Thus basalt is the preferred option. However, to sequester 1 billion tons of CO_2, more than 3 billion tons of basalt would have to be spread: a mindboggling amount equal to almost half of the current global coal production.

soils. This has been discussed in detail over several years and its efficacy remains questionable, while its deployment at scale would again require large plantations and infrastructure (Paul, 2011). Other ideas include adding nutrients to the oceans to encourage plankton to bloom and then sink, carrying CO_2 with them for unknown periods of time—this is not necessarily a permanent sink.[8] Some have proposed intervening in various ways in ocean currents; or enhancing the upwelling and downwelling of water in different parts of the oceans. This last is meant to pump nutrients to the surface to encourage plankton growth, while the plankton and their CO_2 would then sink back down to the depths. Again these would be, or involve, major engineering projects and there is no real data on the long-term effectiveness of any of these approaches. Moreover (and this is a point we shall return to), it is hard to see how there *could* be, without an experiment of such a scale that it would be reckless to begin the experiment in the first place.

By contrast, restoring seagrass meadows and farming seaweed is a safer potentially nature-friendly process that should be investigated swiftly and scaled up (Greiner et al, 2010). Similarly, management-intensive rotational grazing, which mimics the way that flocks grazed before domestication, can increase soil carbon drastically (Machmuller et al, 2015). A key part of our positive claim in this chapter is that agroecological techniques such as these, which do not attempt to manage the climate as a whole through aggressive technological intervention, but rather to *reduce* our malign influence on it so that natural systems and patterns can re-establish themselves, are the alternative to geoengineering, one that should be pursued as the complement to radical emissions reduction.

Many of the geoengineering ideas listed above are based on multiple models and projections developed by many different interests. However, models cannot be relied on (Norman, 2015), since earth systems are complex and the climate is turbulent and involves many factors, many of them unknown, which make prediction notoriously difficult, even impossible. A small variation at a single point can lead to completely different outcomes—the so-called butterfly effect or 'sensitive dependence on initial conditions'. This means that there is really no possible way of reliably predicting the impacts of geoengineering.

We would also need to understand how different applications might interact with each other, since we are increasingly told that we would have to use several at once in order to prevent runaway climate change. It seems

---

8 Using the oceans as a CO_2 sink would inevitably promote ocean acidification, which makes it more difficult for marine organisms with shells or skeletons of calcium carbonate, such as corals, to form and may also dissolve existing shells or skeletons, with potentially disastrous consequences.

inconceivable that (the results of) this multiplicity could be understood in advance. This makes such applications necessarily highly risky.

Small scale tests, models and laboratory experiments cannot tell us what the impacts of geoengineering would be—only full deployment could do that. This is itself a powerful reason for thinking that geoengineering is fundamentally distinct even from other dangerous technologies. It cannot be tested precautionarily but only deployed—recklessly.

There are also serious issues of equity to be considered. Climate change itself tends to impact regions and populations of the global south more seriously while some of the proposed geoengineering techniques would also tend to do this, as could poorly thought-out adaptation approaches, leading to increased inequality. Even afforestation and reforestation could have serious negative consequences if they involve huge plantations of non-native trees, especially on so-called marginal land, or land used by local communities but to which their rights are not recognised by governments. This point is worth exploring further, because it brings out nicely the conceptual distinction crucial to this section between geoengineering on the one hand and large-scale but more bottom-up interventions designed to return the geosphere to a more natural and self-sustaining state, on the other.

*Geo*engineering means just that: the (ultra-hubristic) project of seeking to manage—to engineer, to plan and control top-down—the entire planet, the geosphere. Now, if what we do is grow vast (perhaps genetically-manipulated) forest-monocultures and then burn them and seek to sequester the carbon, that would certainly count as an example of geoengineering. And that is what is being planned; as outlined earlier, it is a little-known and terrifying fact that the Paris targets are premised on exactly that plan—terrifying, especially because there is very little reason indeed to suppose that the plan will work, even on its own terms (Rabinowitz and Simson, 2017). We are gambling our planetary survival on technologies, such as this one, that don't even yet exist.

But the right way to plant trees as a response to the climate threat is very different. It is to seek to restore natural wild ecosystems; to recreate forests that used to be there (albeit slightly tweaking what you seed, to reflect the likely coming temperature changes, etc.). This means our moving away from trying to control ecosystems towards working with them, collaborating with some elements (beneficial predators for example) to keep others such as pests in balance. We create a situation where we have to do less, not a situation where we have to seek to control ever more.

This is not seeking to manage—to engineer—the planet. It is *the opposite*—removing our interferences with natural systems, by taking out (for instance) artificially-created grazing land and returning that land to how it was before we got too big for our boots. It would mean reinforcing and recreating, rather than diminishing, the Amazon rainforest—and every other rainforest and major forest that we can.[9]

And this *is* the fundamental logic of precaution. The logic of the '*via negativa*'; move to doing less rather than always more; seek to de-fragilise systems; switch the burden of proof such that anyone wanting to do something radically new needs to provide evidence that what they propose is safe, rather than our having to provide evidence that what they propose is harmful. It is particularly vital that this burden-shifting is effected, so far as geoengineering is concerned; because, given that geoengineering can only meaningfully be done at the planetary level, there is a real danger that its advocates are going to claim that there is no evidence that what they propose to do is harmful—until they have *done* it, by which time it will be too late to call out their recklessness.

The philosophy of geoengineering and real politics

What are the consequences for real politics of projecting climate-engineering approaches? The moral hazard of deterring action to reduce emissions through constantly promising near-future technical fixes is very real and has to be addressed—see below for some detail on this. There is, in any case, nothing to guarantee that geoengineering will not exacerbate the increasing extremes we face: droughts, heat and floods, together with sea level rise and ocean acidification. In fact it is *likely* that if geoengineering is adopted then extremes will be increased in at least some areas, raising the deeply worrying prospect of competing geoengineering schemes being tried out by different parts of the globe, each of which will have negative effects on others (Nalam et al., 2018; Gass, 2013).

To generalise the point: it is highly risky to intervene in complex and dynamic climate systems we do not understand, and it is recklessly risky, if there is a real alternative path (as we are suggesting there is). Rather than fantasising that we can manage or control the entire future of our planet, of which we have made a particularly bad job in the last generation or two, we should accept living in a world that we can never 'fully' understand or predict—and find effective ways to *reduce* our impact upon that world. This is the logic of precaution.

---

9 E. O. Wilson has created a vision of how it could be done across half the planet in his book *Half-Earth* (Wilson, 2016).

The techno-science behind the current development of climate engineering is deeply flawed, based on assumptions and presumptions of certainty, whereas uncertainty is primary in real ('post-normal') science (Funtowicz and Ravetz, 1993). However, the sense that we will 'have to' deploy geoengineering is gradually hardening in the face of the collective failure to take real preventative action. This failure is of course hardly surprising, due to a particular failure to take action in the industrialised countries with most responsibility for climate forcing emissions. If these made a unilateral commitment to cut emissions swiftly, deeply and verifiably, this might well help to build trust in the international arena. Until this happens it is hard to see how real progress can be made.

We also have a dangerous arrogance about human capability, as our technical capacities increase—we can see (perhaps) how far we have come, even possibly how wrong we have often been, but not how far we still have to go to understand the systems we seek to modify. Geoengineering simply extends the hyper-'Promethean' logic that has got us into this fatal mess (Read, 2016).

The very idea of the 'Anthropocene', at least among its fervent fans,[10] seems to show this—the idea that we have moved from the inadvertent manipulation of earth systems to having the capacity to make deliberate interventions (which presumably means we think we know what we are doing). We might develop technology and seek to deploy it, without the capacity to predict or deal with impacts.

That is a pattern that has occurred before, as we will shortly explain.

The need for the Precautionary Principle

As the geoengineering debate shows, it seems likely that human technical capacity to intervene in complex systems will grow faster than our understanding of them, especially those systems that are inherently turbulent and unpredictable, such as Earth's climate system.

However, if that is the case, will our wisdom increase at a similar rate as we become more dependent upon technologies? As technologies become more powerful, does society have the means, tools and the will to make wise decisions about whether and how to use them, and (above all) how to control their development and deployment by corporate actors?

We need to find a way of examining emerging technologies to try to assess their potential for harm before they are fully developed or deployed. The

10 Such as Mark Lynas: see his book *The God Species* (Lynas, 2011).

Precautionary Principle provides an excellent overarching framework for this discussion.

The most widely-accepted definition of the Precautionary Principle (PP) in international law is that contained in the Rio Declaration (1992):

> Where there are threats of serious or irreversible damage, lack of full scientific certainty shall not be used as a reason for postponing cost-effective measures to prevent environmental degradation.

The PP states that when you are at risk of causing 'serious or irreversible harm', even if you are not sure, then you must step back: for example, when our actions may be causing a possible ecocide then we must take a different path. If there is a route available to us that *doesn't* involve potential serious or irreversible harm then it should be chosen over other routes that may involve such harm.

Contrary to what is sometimes claimed, the PP doesn't prevent or discourage innovation—it *encourages* it. By preventing actions and behaviours that could be dangerous, the Precautionary Principle supports the case for companies (and governments) genuinely to innovate within the constraints set by the possibility of serious or irreversible harm, rather than continue lazily to do something risky. However, despite this truth, there is now considerable pressure from corporate interests to prioritise an 'Innovation' principle over and above the Precautionary principle.[11] Such a 'principle' prioritises human ingenuity over human impacts and proposes that the former can solve problems arising from the latter—and do so profitably as well, *instead* of reducing or avoiding those impacts. This 'Innovation Principle' is a rationale for recklessness (and it may well be used to 'legitimise' geoengineering). The point is that there is an asymmetry here: the Precautionary Principle is designed above all to prevent potentially ruinous scenarios. Hopes invested in inventiveness—and profit—cannot outweigh risks of ruin.

In reality, the Precautionary Principle *already is* an innovation principle.[12] For it is lazy commercial activity, happy to profit from a situation forcing silent risks onto the broader public, that typically wishes to maintain the *status quo* of reckless activity (for example the continued use of lead in petrol, or the continued use of huge amounts of petrol); whereas the Precautionary Principle

---

11 The EU has started to give in to these pressures, highly regrettably: on 13 December 2018, the European Parliament voted for the so-called 'innovation principle' for the first time.

12 How it is, is made clear in great detail in Volume 2 of the report: 'Late lessons from early warnings' (EEA, 2001).

forces them to seek instead to find a safer path, a new innovative route. The need for precaution thus often *drives* innovation.

Consider for example the CLARA report, which concludes that: we should stop forest destruction, restore peatlands, end conversion of grasslands to cropland and restore and expand natural forests (Missing Pathways to 1.5°C, 2018). At the same time we should convert from industrial agriculture to agro-ecology, which would *inter alia* constitute a whole series of innovations. In doing all this we should work closely with indigenous peoples and local communities, including peasant farmers, who still provide some 70% of our food in spite of the expansion and claims of industrial agriculture. 'Innovation' shouldn't be restricted to meaning: reckless high-tech innovation.

It would be ironic (though sadly predictable) if the 'Innovation Principle' were used to 'justify' the reckless roll-out of geoengineering; for this would almost certainly lead to *less* innovation in the vital fields of energy-conservation, renewable energy technology, sustainable and regenerative agriculture, etc.[13] For it would 'license' more business as usual where climate-dangerous GHG emissions are concerned.

PRECAUTION IN UNFCCC ARTICLE 3

'The Parties should take precautionary measures to anticipate, prevent or minimize the causes of climate change and mitigate its adverse effects. Where there are threats of serious or irreversible damage, lack of full scientific certainty should not be used as a reason for postponing such measures, taking into account that policies and measures to deal with climate change should be cost-effective so as to ensure global benefits at the lowest possible cost.'

Questions and values related to geoengineering and precaution

The debate over geoengineering provides us with a vital opportunity to call for the democratic assessment of new technologies and advocate for the precautionary principle into the future, as for instance approaches influenced by Hannah Arendt would have us do.[14] If society is to carry out a meaningful assessment of geoengineering, we must decide what questions to ask. For example, what is the effect on our values and ethics if we believe that we can (both morally and practically) freely alter earth systems to counteract

---

13 As we will discuss in greater detail in the section on 'Moral hazard', below.
14 On which, see for example the work of our Green House colleague Anne Chapman.

the climate forcing we are now knowingly involved in? Or: considering how profoundly we depend on ecological systems that we do not yet understand in any detail, is it ethical or scientifically valid to intervene in those systems in ways that may be irreversible? We must not use the excuse that man-made climate change is already causing irreversible damage to those systems as an excuse to seek to engineer them, if there is available a less reckless route that would return them to a more natural, self-sustaining state.

As we have outlined above:

- In order to understand the impacts of geoengineering (which models and laboratory experiments cannot show us, due to the complex nature of climatic systems and the number of variables involved), we would need full deployment, which could easily have irreversible consequences.
- Certain approaches, such as injecting particles into the stratosphere, would have to be continuously maintained, as halting them would lead to an extremely rapid increase in temperature. Is it ethical to oblige future generations to do this? *A highly-climate-stressed future is hardly the safest environment in which to rely on the organisational and resource capacity to maintain geoengineering efforts* such as SRM.

Moral hazard issues

While we play with the idea of geoengineering the planet in order to tackle global overheat, we are not reducing emissions. It seems likely that as long as politicians and many publics feel they can 'change the subject' they will continue to do so, which means that the focus is shifted away from effective action. Geoengineering is a diversion from the real issues, which involve: leaving fossil fuels in the ground, changing our diets, adapting transformationally to the dangerous climate change that is coming—and thereby increasing, not reducing, human happiness.

When it comes to practical questions for the technology assessment process we need to ask many things about the safety and effectiveness of geoengineering and the potential unintended consequences. Above all we need to ask who makes the decisions and how. How does the application of the precautionary principle change this debate? Here is one possible way:

We ... need to find a way of examining emerging technologies to assess their potential for harm before they are fully developed or deployed.

This requires a process based on precaution and work with a wide range of people including scientists, sociologists, philosophers, politicians, and the general public. It is particularly important to consult with Indigenous Peoples and local communities who have their own knowledge systems and cultural references.

— Steinbrecher and Paul (2017: 45)

Precaution and development

The proper application of Precaution actually indicates a completely different development path from that which we are currently following. Instead of the impossible paradigm of endless economic growth we need to shift to a paradigm of economic and ecological justice, which means contracting and converging—the rich give up privileges and the very poor gain some—and all this must take place within planetary boundaries, that is, within the capacity of the planet and the ecosystem functions essential to our lives.[15] As outlined above, precaution also means a different approach to the development and deployment of new technologies which is often described as blocking innovation but which can actually encourage a new kind of innovation, less risky, less short-term in focus. This too will be a part of the required new development path.

An objection?

Some will say however that the days of precaution in relation to *climate* are over, because the evidence is in, the science is settled.

It is true that evidence-based (climate-) science now clearly provides a sufficient basis for the need for radical action to address the changing climate, and in this way climate differs from other threats (for example genetically modified organisms, or GMOs) against which the evidence-based case alone is inadequate (Read, 2015; Taleb et al, 2014).[16] (The evidence against GM is

---

15 On which, see Kate Raworth's book *Doughnut Economics* (Raworth, 2017).

16 Climate-change deniers downplay the risk of human intervention in natural systems. GMO proponents similarly downplay the risk of human intervention on natural systems. GMO proponents are in an analogous position, therefore, to geoengineering proponents. In both cases, rampant technophilia runs beyond any evidence-base, and there is no evidence-based case for the safety of the technology in question. (Our thinking in this note is influenced by Joe Norman.)

not overwhelming, in the way that the evidence against climate-change denial *is* overwhelming. What is overwhelming against GM is the purely precautionary case: which translates into saying that the evidence-based case *for* the safety of GM crops is far from adequate.)

However, the Precautionary Principle remains relevant and important in the case of climate generally, and specifically of geoengineering. Examples of its general relevance include:

I. 'Climate-sceptics' emphasise the uncertainties inherent in many aspects of climate science. They are right. What they have not understood is that *uncertainty* makes the argument for climate-action stronger, not weaker. For uncertainties cut both ways. If we are uncertain where tipping points are, how different tipping points may interact, or what the level of 'climate-sensitivity' is, then we should be *more* precautionary, not less, because the outcome is harder to control, and could be even worse than we predict. It is cherry-picking to assume that uncertainty always points in the direction of climate science being 'alarmist'. The reverse may well be true.[17]

II. There remains significant uncertainty about key elements of climate science as these are applied to what is required for a good quality of human life in the long term. For example: is 1.5 degrees (let alone 2 degrees) really a 'safe' level of heating from pre-industrial temperatures, or might even 1.5 degrees lead long-term to complete break-up of the Earth's ice-sheets, or render significant parts of the Earth's surface long-term-incapable of supporting significant food-production for human consumption (let alone for supporting a wide diversity of species including humans)? We dare not await complete answers to such questions before acting precautionarily. The PP enjoins us always to err literally on the safe side. It keeps us safer than a purely modeling- and risk-based approach would.

III. Looking back over the last generation, especially the last few years, there has been a persistent tendency for various outcomes to *exceed* the 'likely worst-case scenarios' of climate modellers. Consider the unprecedented Arctic temperatures and diminishing sea ice recorded recently. This would have been less of a disaster, were we already taking strong precautionary action. The same is true, going forward.

Now, how do these three points impact the case for geoengineering?

---

17 Thus the Precautionary Principle offers an independent argument for strong action on climate, even to those unconvinced by climate-science. It has the capacity to persuade, on the basis that even a low probability of the climate science being right would already demand that we act strongly and precautionarily because, if it is right, we risk catastrophe.

The idea of accepting raised levels of CO_2 emissions, and seeking to engineer the workings of the planet to avoid the harmful consequences of those raised temperatures will, we have no doubt, become increasingly popular in the next few years, as we face increasingly the results of our accumulated emissions, our accumulated climate-recklessness and refusal to take action. It is clear that we have failed to listen sufficiently to climate science, and failed to act precautionarily, and are moving inexorably and fairly swiftly closer to the dangerous 1.5 degree temperature-rise threshold. Our argument has been that *geo-engineering is a highly reckless response to this situation*, not a precautionary one. We have shown that by definition there can be little that would constitute empirical evidence that geoengineering is a bad idea before it takes place, since it can only really work at a planetary level and therefore cannot be introduced in small-scale experiments. This is why we should apply the Precautionary Principle: We ought to seek a route to climate safety that does not rely on an untried and hazardous experiment with the whole Earth. We need instead *to put an 'emergency brake' on emissions.* And, given that that is tragically unlikely to happen, we need, as Green House have been urging, a programme of transformational adaptation, and indeed a series of measures taken at all levels to prepare for possible collapse and to start living now on much less.

The case for geoengineering is reckless. There is a strong precautionary case *against* climate-engineering.

Taking stock

To sum up the argument so far: the situation is genuinely desperate. That desperation is being used to argue that geoengineering is required in order for us to save ourselves. But: we are in fact *choosing* to fail at present. The real question is 'only': how badly are we going to choose to fail? The difference between 'badly' and 'atrociously' is a big difference, in terms of harm-reduction. We in Green House hope that we will choose to fail badly, not atrociously.

However, geoengineering gives us a tacit 'excuse' for failing atrociously. The extreme moral hazard that it leads to will 'legitimise' a cocktail of massive ongoing GHG emissions and reckless rolling out of untried, costly, hazardous technologies.

The future will be bad. But we don't have to behave atrociously and risk everything, as continued high-emissions pathways and geoengineering do. We can instead choose to seek to do the best we can to create a future which is as 'least bad' as possible. What does this mean?

Collective decisions to consume less (whether made by enlightened cities, or rural areas, or particular groups of people voluntarily) would be a true expression of values and democracy. Many indigenous groups take time to make decisions, partly because they believe that everyone must be included and reach a consensus on the final decision. Perhaps we should learn from them if we are to respond adequately. (Probably we *will* do so sooner or later, and we could try to make it sooner.) Our so-called democracies give no space for real deliberation by the people over the kinds of issue central to this chapter. What is needed for it to be possible for the global community to eventually agree to take real action? Real information, time, trust... However, elected governments are always focused on staying in power, and therefore they often do not represent the true, long-term interests of the people. Moreover, they typically lack any mechanisms for representing future people.[18]

Thus below we propose a major shift in the way that deliberation and consultation take place, in relation to new technologies. The application of precautionary thinking strongly suggests that instead of going for novel top-down technologies whose impacts we cannot predict, we should prioritise the actions whose impacts we can basically understand—for example reducing emissions, halting forest, ecosystem, soil and water supply destruction and degradation and reforesting sensitively with native trees. We should 'deconstruct' our impacts rather than build them up. We should simplify the aspects of the system involving and dependent on us and our agency, rather than complexifying that system further. We should not fragilise the Earth system further—and geoengineering inevitably fragilises, because it complexifies yet further and thus builds in additional ways in which things can go wrong; for instance as a result of the need to maintain geoengineering once it has started. As noted above: *It is utterly reckless to depend on geoengineering to be long-term sustainable, given that we are uncertain that we can sustain the operation of industrial civilisation at a high level of inputs.*

How can we start to shift away from the current climate of opinion, in which geoengineering risks coming increasingly to seem 'necessary'? We should turn for inspiration to indigenous peoples, and adopt an approach which means thinking about the welfare of the next seven (or indeed 77, or 777) generations before making such major decisions. This would entail tackling the chronic short-termism of currently-hegemonic modes of politics, technology and development, and thinking beyond short-term human advantage to properly consider the biosphere on which all lives depend.

---

18 On which, see Rupert Read's 'Guardians for future generations' proposal for Green House.

Facing up to climate reality surely requires not that we go in for geoengineering but that we refrain from doing so, if only out of humble awareness that climate reality itself may scupper the long-term viability of sustaining geoengineering programmes.[19]

However, the mainstream is a very long way from accepting any of this. In fact, we have a lethal combination to deal with: the profit imperative, the 'normalcy bias' and the fascination with technical fixes.

The proper application of precaution could address this too. We know we have to take action, but we do not seem prepared to take the right action. Instead we keep going for diversions, because the right action means fundamentally rethinking our model of development starting with those countries that have followed that model for the longest time. Such fundamental rethinking may eventually come—once enough disasters have shown the utter inefficacy of the current model,[20] and have started to take down growth-obsessed international capitalism.

What is needed is not extreme technology, but rather action at every level by people, not just government but everybody, starting with those who are not too poor or hungry to act. Governments need to communicate with people to explain what is needed and help to provide an enabling context.

Precaution can completely change the nature of the debate and take it out of the hands of the technocrats. What we need is often low-tech and involves the cooperation of networks of people at ground level, not top-down solutions from government and big corporations.

We turn now to thinking about how society and the economy could be differently organised, so as to change this dangerous dynamic.

The stage gate process

There are often several phases in the development of a product or technology and the aim of a *stage gate process* is to identify points where the proposed development should be examined and a decision taken as to whether or not to proceed to the next stage. It was in fact (as we will discuss below) a stage gate process that helped to stop the first proposed geoengineering project in the UK, the SPICE (Stratospheric Particle Injection for Climate Engineering)

---

19 Instead we should do as much as we can by way of mitigation and transformational adaptation – while fully aware that it is unlikely to be enough to head off bad outcomes, disasters.

20 This is the hopeful argument of the chapter in this book authored by Rupert Read and Kristen Steele.

project (in 2010), which was to use a tethered balloon and hose to disperse water at a height of 1km to try out a prototype for delivering particles into the stratosphere at some 20 km above the surface of the earth. In the stage gate process for this project:

> ...a panel of external experts considered the progress of the project against a number of criteria, such as checking that mechanisms have been identified to understand wider public and stakeholder views on the envisaged applications and impacts.
>
> Following the stage gate meeting, the panel advised the research councils and the SPICE team that further work on stakeholder engagement and the social and ethical implications was required.[21]

In order to properly assess geoengineering we need a thorough stage gate process combined with the strict application of precaution and ongoing public consultation. Public consultation should happen at every stage of the development of a new technique and should have the power to halt it completely. If such an approach had been taken to polychlorinated biphenyls (PCBs) and asbestos, it is possible that their development and deployment could have been halted early in the twentieth century (EEA 2001).

Public consultation

This is a vital part of any precautionary process of technology assessment. Representatives of the public, randomly selected to have no particular bias or knowledge of the subject, have consistently shown themselves to be perhaps-surprisingly wise advisers. They are not specialists, but nor do they have specific *interests* and they have a *broader perspective* than any group of specialists.

However, there has so far been too little public consultation on the subject of geoengineering and certainly no attempt to set up a continuous process of consultation to follow developments as they take place. One example from the UK provides some key insights into the principles that should be applied to any discussion of geoengineering. In 2010, the Natural Environment Research Council (NERC) held a public dialogue on geoengineering in the UK.[22] The members of the public involved came to some strong basic conclusions, relating to human ignorance of climatic systems, justice, and equity. These

---

21 See https://www.epsrc.ac.uk/newsevents/news/spiceprojectupdate/.
22 See https://nerc.ukri.org/about/whatwedo/engage/engagement/geoengineering/.

are highly relevant and make an excellent starting point for any discussion of geoengineering today.

NERC CONSULTATION ON GEOENGINEERING: ETHICAL IMPLICATIONS

The members of the public concluded that:

- We have no right to interfere in complex ecosystems if we do not understand what we are doing, or the consequences, or if (and in what ways) the impacts will last a long time.
- The values involved go beyond economics to include social and ecological values.
- The rich do not own the planet and have no right to exploit it for gain or increased inequity.
- As much of the population of the world as possible should be included in making decisions that will affect them (like on geoengineering).
- The UK population should be given as much information as possible to enable them to participate in making such decisions.
- Scientists need a public mandate to move forward with any geoengineering.
- The public that considers whether to give them that mandate should be given as much information as possible in order to decide.
- Interfering with natural systems using geoengineering could legitimise further interference later, dubiously.

As regards uncertainty of outcomes, the public typically takes a pretty full precautionary approach to the consequences of intervening in complex and delicate planetary ecosystems that we do not understand. Ongoing public consultations could be part of a required, rigorous, non-commercial, scientific stage gate process, that is an examination of progress and questions that arise regarding the investigation of a proposed application of geoengineering at every stage of its development, with the mandate to halt further development at any point.

Our guess is that, if stage gate processes were properly deployed, geoengineering never would be.

Conclusions

It is good to discuss geoengineering because it reveals what a dangerous situation we are in, as regards the climate and especially human hubris. But it is one thing to talk about geoengineering and quite another to recommend implementing it. We recommend *talking* about it—so that the enormity of what the proposals reveal about us and the situation we have got ourselves into can hopefully wake us up so that we don't actually *do* it.

In fact the sheer arrogance of geoengineering proposals reveal just how deluded we have become about the power of our technologies. Anyone who believes that they can successfully engineer an interacting series of complex systems that we mostly do not understand and which are innately unpredictable and incredibly powerful must be either deluded, or, at minimum, recklessly over-optimistic (especially given 'our' record to date). What we need now is a strong dose of humility and a recognition that all human inhabitants of the planet are in this together and need to collaborate respectfully, learning from each other—including via the kind of inspiration from indigeneity, and the more prosaic but very valuable possible deliberation methods, described above. We must recognise that we depend for our lives on these planetary systems that we now contemplate altering deliberately. At the same time we urgently need to cease destroying ecosystems and their human and non-human inhabitants, and disrupting climate systems at every level.

This implies a philosophical shift, away from arrogance about human technical capacity to resolve any problem we create, and to do so with a main eye on profit, towards a better sense of epistemic humility and our proper place in the planetary system. We are used to believing in 'progress' and assuming that technology represents 'innovation' and therefore 'progress', while indigenous and biocultural knowledge of all kinds allegedly belong to the past and must be superseded. What we need to do now is to strongly question this position. Doing so would represent a true paradigm shift away from the mindset that conjures up geoengineering.

Climate-realism enjoins accepting that the human race is walking more or less knowingly into disaster, but nevertheless seeking to prevent catastrophe (On this, see John Foster's chapter). There will be bad climate-damage; there will be disasters. (See the chapter by Rupert Read and Kristen Steele, on the lesson of this.) Geoengineering pretends that we can avoid disaster while continuing with a model of development based on high emissions, but it could actually contribute to a catastrophic situation: one in which there are huge ongoing GHG emissions *and* reckless rolling out of geoengineering technologies.

Geoengineering is both risky in itself and a dangerous diversion from what we should be doing, urgently, as a global community: reducing emissions and sensitively restoring the ecosystems that are our life support. This process must be honestly led by those with most responsibility for global overheating: the nations that were the first to industrialise. Our technical capacity has outrun our ethical frameworks and we urgently need to focus on strengthening the latter and applying the Precautionary Principle. We have plenty of clear examples of how the emergence of new technologies highlights the assumed positive aspects long before the negative impacts become clear—except to a few, but they are often disbelieved (just as the prophet Cassandra was doomed to prophesy truthfully and never to be believed). But those who cannot learn from history often have to repeat it, as examples such as the development of PCBs show. We must try to avoid repeating this kind of mistake, because the stakes are now too high. There are fewer unknown risks in reducing consumption of energy and resources than there are in deploying untried technologies, many of which would have to be maintained beyond the foreseeable future and whose potential impacts and interactions cannot be fully known until they are deployed. Our legacy to future generations does not look like a happy one. We need to change. Given new political will and commercial frameworks, this may still just be possible, but time is rapidly running out.

So that is our pitch. Climate-reality is going to hit us hard, whether or not we geoengineer. Some of our ideas in this piece may seem politically unrealistic at present, and they probably are. But *they will come to seem realistic in time.* The only thing that might stop them from doing so is if we decide to send ourselves to sleep again by deluding ourselves that 'negative emissions technologies' can save us and allow a continuation of near business-as-usual. The real danger of geoengineering is that it is a continuation and indeed accentuation of *the very mindset*—'progressive', reckless, anti-revolutionary, basically mindless, while stuck within human solipsism, and without respect or love for the Earth—that has set us on our current tragic path. As such, it could prolong that path beyond a point of true no-return, a point of runaway climate-damage *or* of catastrophe induced directly by geoengineering technologies (for example, a catastrophic failure of the world food system, if BECCS were rolled out across an area almost twice the size of India, which is what it would need in order to be effective; or a catastrophic failure of the world's weather, if for instance we lost the monsoon completely, as a result of deploying SRM).

We will take an alternative route. That route will either quite simply be forced upon us, by collapse; or (one hopes) will come to seem realistic before that point—when we finally are willing to face up to climate reality.

Acknowledgements

Thanks to Tim O'Riordan, Ray Cunningham, Anne Chapman, John Foster and Naomi Marghaleet of 'You Said It' Ltd for comments.

Part III: Framings

Chapter 7

WHAT THE CRISIS OF THE LATE MIDDLE AGES IN EUROPE CAN TELL US ABOUT GLOBAL CLIMATE CHANGE

Brian Heatley

The thing that hath been, it is that which shall be; and that which is done is that which shall be done: and there is no new thing under the sun.

— *Ecclesiastes* I, 9

Introduction

History has been affected by climate change before. Climate change, and wider environmental degradation, may have played a part in the fall of the Roman Empire, though the evidence is scant (Ponting 1991, 78). A better documented case can be made for the events of the Late Middle Ages in Western Europe, very roughly 1300–1500 CE,[1] and long characterised by historians as 'The Crisis of the Late Middle Ages' (Fossier 1986, 1–14). A time of famine, plague, peasant revolt and wars, during which the European population dropped by half. This catastrophe followed three centuries, the High Middle Ages, of expanding population, economic growth and intellectual progress—a bit like the 300 years from 1700 to 2000. Many other changes took place in the High Middle Ages: the ideas of religion were reconciled with the rediscovered ancient philosophers, feudal bonds began to be weakened, while technology, towns, industry and the nation state all took small steps towards the modern age. But Europe's population had reached its ecological limits, and the climate got a little colder.

---

1 All dates in this piece are CE.

The purpose of this chapter is to compare what happened in the Late Middle Ages, that is roughly 1300 to 1500 in Western Europe with what faces the whole world in the next two centuries. The two periods, and the geographical areas for comparison are of course radically different in scale and in nature. But, as we will see, two of the most important features of Europe in 1300 and the world in 2000 are the same; over-population and broad ecological crisis combined with climate change, albeit today the world faces significant warming while medieval Europe faced just a little cooling.

We begin by saying how we got to the two starting points in 1300 and 2000.

The prelude: 300 years of progress

Most readers will be broadly familiar with what has happened in the 300 years between 1700 and 2000. In short, there have been the scientific and industrial revolutions, and world population has grown massively. In some parts of the world, the rich countries, most people have led longer and materially better lives, with growth in democracy and an improvement in the position of women. Religion, challenged by science, especially evolutionary biology, has often been replaced by more secular outlooks. Meanwhile poorer countries, often colonised for most of this period by the rich ones, have not done so well, and many still live in abject poverty. There have been major wars between the great powers, but also long periods without them such as much of the nineteenth century 1815–1914 and from 1945 until the present. But also there has been great environmental change, with extension of the area under cultivation, more and more people living in towns and cities, and pollution, climate change and over-exploitation of many natural resources.

The history of 1000 to 1300 in Europe, called the High Middle Ages by historians (Mundy 1991), is less familiar. As in the last 300 years, the population grew, food production rose, and towns were established and grew. And, as more recently, the money as against the subsistence economy also grew. The feudal system, where peasants held land from their lord in exchange for labour on the lord's land, was giving way gradually and patchily to a system where peasants simply paid a money rent. But unlike the modern era, this expansion was not based on technological change. There were some innovations—the wider adoption of a heavier plough pulled by a team of horses was perhaps the most significant as it allowed the exploitation of heavier soils—but the increase in food production, and the larger population, both rural and the urban population it supported, derived almost entirely from simply increasing the area of land used to grow crops. Most of the new land came from clearing woodland, and by 1300 in most of Western Europe, there were no more available substantial woodlands left to clear.

Life in the High Middle Ages was dominated by the Roman Catholic Church. No modern institution matches its power and institutional reach. The Church extended its rule over ideas, education, practical personal morality and everyday life, and the legitimacy of rulers. The modern state did not exist; great nobles held land in many different geographical areas with no particular cultural or economic coherence, and their quarrels were usually family and dynastic rather than anything to do with what might be called 'English interests' for example. In the High Middle Ages a dynamic Church created many new institutions, including the great Gothic cathedrals of Europe, and the first universities.

While there was no scientific or technical revolution in the High Middle Ages, there was an intellectual revolution. Christian theology was challenged through contacts with the Arab world after 1000, and by the rediscovery of ancient learning, and pre-eminently by the works of Aristotle. For some time the church reacted by suppressing ancient knowledge, but eventually Thomas Aquinas created a synthesis of Christian and ancient ideas which has ruled within the Catholic Church to this day (Mundy 1991, 327).

All in all the European world of 1300 looked very different from 1000.

1300 and 2000

Before going on, I think it worth drawing out here the similarities and differences between Western Europe in 1300 and the world in 2000. There are three major similarities—that in both cases the economy had or has just about reached its ecological limits, that in both cases (but for different reasons and to different degrees) it faced or faces climate change, and that in both cases there was and is a curious intellectual complacency about the nature of the future.

The major differences include the fact that today we know about climate change, which is but an example of our scientific knowledge and technological awareness, that we are aware of the possibility of technological innovation and its role in our lives, the fact that our expansion has been much greater and some of us are much much richer, and that our arrival at ecological limits is both global and so much more comprehensive.

Similarities: environmental limits

I am not going to reproduce here the evidence that our world and economy in the early 21st century has reached its ecological limits. Perhaps the best summary of the evidence for this is that produced by the Stockholm Resilience Centre's comprehensive research on Planetary Boundaries. This identifies 9 boundaries:

1. Climate change
2. Change in biosphere integrity (biodiversity loss and species extinction)
3. Stratospheric ozone depletion
4. Ocean acidification
5. Biogeochemical flows (phosphorus and nitrogen cycles)
6. Land-system change (for example deforestation)
7. Freshwater use
8. Atmospheric aerosol loading (microscopic particles in the atmosphere that affect climate and living organisms)
9. Introduction of novel entities (e.g. organic pollutants, radioactive materials, nanomaterials, and micro-plastics).

The research then cites evidence to show that human action has already breached four of these boundaries, climate change, biosphere integrity, biogeochemical flows and land system change (Steffan 2015). By contrast, in Europe 1300 there was essentially only one problem, the shortage of further arable land in relation to the size of the population, or, in our modern jargon, land system change.

The evidence that there was no further room for agricultural expansion by around 1300 is hard to find; there are no readily available consistent statistics of land usage from this time. The extent of land shortage would anyway have been patchy. But a broad picture of the ability of the land to feed the population can be illustrated from the average heights of people buried at different times; shorter people meant less food. There is evidence that average diets in England for example became less satisfactory from as early as 1200. The decline in height, because of poorer diets begins in 1200, not 1300, suggesting that population pressure on a finite supply of land was a factor all through the century. There is evidence (Galofré-Vilà 2017, 31) that average diets in England for example became less satisfactory from as early as 1200. A decline in average height, because of poorer diets begins in 1200, not 1300, suggesting that population pressure on a finite supply of land was a factor all through the century.

Similarities: climate change

As with the environmental crisis, I will not rehearse the evidence to show that even with the Paris Agreement we now face climate change in the range of at least a 3–4°C global temperature rise by 2100, and more thereafter. Even the UN now estimate that we are actually on track for global warming of up to 3.4°C on the basis of the Paris Agreement being implemented (UNEP 2016, xi)—which the author of this report predicted as the Paris Agreement was reached in late 2015[2] (Heatley, 2015).

The Late Middle Ages was preceded by the Medieval Warm Period (Wikipedia 2018a). While it is hard to be precise about the boundaries, this is generally held to have occurred in the 300 years between about 950 and 1250. The following colder period is normally called The Little Ice Age, generally held to have happened between 1300 and around 1850 (Wikipedia 2018b). In both cases this was not a global phenomenon, but restricted in the case of the Medieval Warm Period to the North Atlantic region, and perhaps more widely in the Northern Hemisphere in the case of the Little Ice Age. In both cases the evidence is largely indirect and unsystematic, and includes evidence from botany (e.g.. tree rings, evidence as to what plants grew where) and documentary records.

However, it is clear that the average temperature difference between these periods both globally and in the whole of the Northern hemisphere was modest. The IPCC Third Assessment Report 'show(s) temperatures from the 11th to 14th centuries to be about 0.2°C warmer than those from the 15th to 19th centuries, but rather below mid-20th century temperatures' (IPCC 2001, para 2.3.3). A summary of 11 recent studies (many of which have been contested) of reconstructed average temperatures over the past 1000 years suggests that the greatest apparent dip in average temperature between say the mid 1100s and around 1450 is still only about 0.5°C (Wikipedia 2016). It looks likely therefore that the actual change in Western Europe from around 1300 to 1500 was a very much smaller average temperature loss than the 3–4°C rise we are now expecting to occur globally. We have rather little knowledge of what form it took, how far rainfall changed, and whether the incidence of particular types of weather, such as severe storms altered.

We can conclude only that the climate change experienced during the Crisis of the Middle Ages was probably modest compared to what we face now, and of course in the opposite direction, colder rather than warmer.

Similarities: ideas

My final similarity is perhaps more controversial—the idea that both in 1300 and 2000 the prevailing elite attitudes were in both cases curiously intellectually complacent about their impending futures, and that the dominant

---

2 And see also the introduction to this volume.

intellectual practices, scholasticism and neoliberal economics, show considerable similarities.

Thomas Aquinas's detailed synthesis of Aristotle and Christian theology meant that around 1300 an educated person could think that the question of human knowledge was essentially finished and completed, so that there wasn't anything major more to know. Moreover, and unlike now, a few highly educated people could reasonably claim to have mastered personally the whole of human knowledge. So early in the fourteenth century an educated man like Dante (1265–1321) could draw on this synthesis to write his *Divine Comedy* (Dante 2003), a trip through a very Christian Purgatory, Hell and Heaven, with most known characters from the ancient classical world safely fitted into the Christian framework. True, some intellectual dissenters existed or would shortly arise (Roger Bacon, Wycliffe, William of Ockham and Jan Huss for example) but apart from Huss their wider influence was slight, and the extent of their influence has often been exaggerated, as modern authors have sought ever earlier origins for the emergence of the scientific method or the Reformation.

In a way that situation is profoundly unlike the one we have now. We expect knowledge, especially scientific knowledge, to expand continually. Many believe that technology in particular will solve all our contemporary environmental problems. But I would argue that this is actually more similar to the Medieval position than we might think; in 1300 knowledge was at a level that was constant, while in 2000 we regard its *continual expansion* as a constant, a new normal, not ever expecting that the great apparatus of scientific research or technological development will ever halt or falter, or be unable to solve our problems. It is a kind of complacency, but with a different subject.

Moreover, outside science and technology, there is, in the West, an overwhelming consensus about how broadly to run political affairs, through religiously tolerant, democratic nation states, with market economies and high levels of individual freedom. For the rest of the world the test of modernisation is the extent to which they accept this broad idea. Yet this dominant consensus is intellectually short-sighted, unable for example to respond to what its science tells us about climate change. Similarly there was in 1300 a consensus about how a good ruler should behave. While the consensus was different, it was at both times a consensus, which cannot be said about times of political and intellectual flux like for example 1500 and 1900.

Finally, in the sphere of more obviously elite intellectual activity, there is one other parallel worth exploring. The most learned people around 1300, all in practice churchmen, operated within a system of thought known as scholasticism (Verger 1986, 160–165). This created a great edifice of logical thought, that, drawing on both revealed Christian religion, and the older classical philosophical tradition, provided explanations for every kind of situation and problem. Scholars made their reputations from arcane disputations and learned deductions within this great logical edifice, where controversial positions could be debated and tested, with remarkably little connection with the real world. But the real world of politics and government listened and believed and applied the conclusions.

There is a modern parallel to this (explored in Rapley 2017). In our contemporary political life, especially since the 1980s and in the UK and US, a constant touchstone is the pronouncements of a particular academic group, the neoliberal economists. 'Any interference with the free market will lead to a sub-optimal result' we are told. This and similar propositions, which are really only the more modern manifestation of a longer tradition of laissez-faire economics which held sway through most of the nineteenth until the mid-twentieth century, are derived from a complex system of axiomatic mathematics. By assuming certain very limited axioms about human behaviour, a whole system of logical deduction is developed, which purportedly proves variations on the theme that provided all economic agents are free, including allowing a few to become very rich, and many to remain very poor, all will be for the best in the best of all possible worlds (Lipsey 1999, 287). Economists' reputations depend upon their skill in making deductions within this axiomatic world, not upon their abilities to explain the real one—the profession seemed little abashed by its inability to predict the 2008 financial crash. There is in fact a lot wrong with both the assumptions, the deductions and the definition of 'best' (see for example Keen [2001], or Shaikh [2016]) but that's not the point I want to make here.

The point I would stress is the logical, social and functional similarity between medieval scholasticism and modern neoliberal economics. Both logically purport to derive theories of great practical political importance for the real world from plausible (in the context of the two very different relevant times) but incorrect and partial assumptions. Both are intellectually difficult, abstruse and arcane. Both are carried out by high status individuals, almost always male. Both served and serve the interests of existing elites in their respective societies and have largely conservative implications for politics and society. Scholasticism lingers on in the Vatican but now has little practical significance; what will be the fate of neoliberal economics faced with a world where resources are not limitless and externalities dominate material life?

Differences: knowledge and the scale of the environmental problem

The difference between our situation now and 1300 are all of course but aspects of the similarities outlined in the previous section.

The first and most obvious point is that we have more awareness of our situation. We know climate change is coming and why, and we have some idea of its effects. We have thought ever since Thomas Malthus in the early nineteenth century about the consequences of ever-expanding population upon a finite earth (Malthus 1798). We (mostly in Europe, but less elsewhere) no longer think of these things as the result of the will of God, either incomprehensible, or as a punishment for our sins. We have a far better and more accurate grasp of history and past similar events. This gives us the power to do something about our predicament in a way that was impossible in 1300.

This optimistic point needs to be balanced by a far more pessimistic one. While north western Europe was in 1300 on the edge of a population catastrophe due to having out-grown its available agricultural land, this was essentially its only environmental problem, the only one of the planetary boundaries it had crossed. Once the population pressure was relieved by about 1400, life improved considerably until the same problem began to repeat itself in the early modern period from 1500. Our problems in 2000 are far worse. First the climate change we face is an order of magnitude greater than that faced in 1300. Second, we face a global crisis, not one largely confined to one part of the world. Third, besides land use and climate change, we face other global environmental crises, including pollution, disruption of the phosphorous and nitrogen cycles and bio-diversity loss.

Material life 1300–1500

Food and population

We know little about absolute population levels in the Middle Ages, but the figures were clearly very low by modern standards, at best a million or two in England, and no more than tens of millions in Western Europe overall. The very rough story that we will outline here is that population almost doubled between 1000 and 1300, was reasonably steady overall up to the Black Death around 1350, after which population fell in the last third of the fourteenth century by perhaps a third to a half. Then the population recovered to something like the 1300 figure by about 1500 (Wikipedia 2018c, especially where citing Urlanis, for an account that purports to far higher accuracy than is reasonable.)

We have seen that by 1300 western Europe was having trouble feeding itself, having run out of agricultural land, and with no technological changes in agriculture likely to increase productivity per acre. In 1315–17 came the first crisis, a great famine, precipitated by bad weather in 1315 and crop failures until 1317. It was sometime before food supplies were restored, and in many areas between 10 and 15% of the population died.

Famines were repeated in 1330–34 and in four subsequent periods in the fourteenth century. Many modern famines are often rather less an absolute shortage of food and more a failure of the economics of food distribution; those who die are too poor to buy food that is actually available on wider markets. While no doubt this was partially true in the Middle Ages, particularly for the relatively low number of vulnerable urban poor, in practice transport constraints were far more significant. Famine was often very local, and the close association of the relatively huge numbers of medieval peasants with food production meant they were in a better position than modern peasants to keep practical control of what they produced.

But more significant for mortality than famine itself was the arrival of the Black Death in 1346, which moved across and persisted in Western Europe until 1353. The Black Death was probably Bubonic Plague, and it killed in many areas as many as one third of the population. It returned in subsequent episodes in the fourteenth and fifteenth centuries. Although plague was the immediate cause of death, its effects were as drastic as they were because it was acting on a population much weakened by famine that had its origins in over—population and poor harvests caused by climate change. The recurrent wars and peasant unrest of the period added to the overall mortality.

After 1400 the situation for the survivors begins to improve. The over-population had been reduced, and land abandoned during the later fourteenth century was brought back into cultivation. Even though the weather got no better, overall food supplies were more robust, and although the plague was to persist until the late seventeenth century, population levels began gradually to improve. But by 1500 over-population began again to be a problem, and living standards began to fall again, only to recover as agricultural techniques improved and Europe was able to draw on the fresh ecological resources of the Americas, initially by emigration and later by food imports.

Globally we face a rather similar position now, a population running ahead of our agricultural resources, and rather more serious climate change. Setting aside for the moment the possibility that we may find global technological solutions, millions of deaths are likely to arise directly from climate change. These include deaths from extreme humid heat in certain regions near the equator (*Guardian* 2017) and deaths from the direct effects of extreme rainfall, storms and flooding.

But awful as these direct effects of climate change are likely to be, they are minor compared to the indirect effects of famine, war and epidemics. We are already seeing, in the Sahel, famine whose main cause is almost certainly anthropogenic climate change. The experience of the Great Famine of 1315–17 shows how a population already at ecological limits and malnourished can suffer mortality of 10%. Moreover, in the Middle Ages the land had not been compromised, it was simply over-crowded. Once the population halved, it could easily feed those who were left, and for a while to a better standard than before. Now climate change will degrade the potential of land, by disrupting weather patterns and water supply, making traditional agricultural systems no longer appropriate. Our famines may be more prolonged as both the demand for food but also its supply are diminished.

Epidemics though seem different; it is not obvious that we will necessarily face an epidemic like the Black Death, which caused most of the population decline from 1350. We, especially in the rich world, have assumed in the years following World War Two and the early triumphs of antibiotics, that epidemics are entirely a thing of the past. Surely faced with a modern epidemic we would have far greater defences, including better knowledge of disease transmission, better ways of combatting particular illnesses, vaccinations and knowledge of how populations can behave in ways that will help them. Much of the ability to resist epidemics is not specifically medical, or even very technologically complex, it is often the simple ability to isolate infected people, maintain good water supplies, and deal effectively with sewage.

But the notion that epidemics are a thing of the past is probably an unwise assumption (MacNeil 1976). The HIV/AIDS epidemic, and recent Ebola outbreaks in Africa have shown how dangerous modern epidemics can be. Our technical advantages are very unevenly spread; most rich countries have faced no serious epidemic since the flu epidemic following World War One, and have generally coped well when epidemics have arisen. Poor countries have fared less well and the poorest of the world would probably do no better against a serious epidemic than did the Europeans of the 1350s.

Indeed, as populations weaken due to reduced nutrition, and medical services and social controls collapse, a major driver of human mortality is likely to be epidemics of one kind or another. The increasing resistance of many disease organisms to antibiotics is well known. The effects of epidemics will typically be patchy, almost at random some areas will suffer terribly while other similar areas will survive. Moreover, an increase in one of the most damaging diseases, malaria, is likely simply because the range of the relevant disease-carrying mosquitoes will increase. Also, unlike in the fourteenth century, disease will travel quickly over vast distances. In short the lesson from the Late Middle Ages seems to be that large numbers will not die of climate change as such, or even directly of its consequences for food supplies, but more likely from epidemic diseases.

What does the experience of the late Middle Ages have to say about the scale and rate of mortality? Although population doubled between 1000 and 1300, that was nevertheless a very slow rate of growth by modern standards, being around 0.2% pa. By comparison, world population is growing now at about 1% pa, which by comparison means a doubling every 50–60 years. The population halved in many regions in broadly the last half of the fourteenth century. But a population decline by half over fifty years is not huge in annual terms—it's simply a 1% pa decline.

The point for comparison between now and the Crisis of the Middle Ages is just how muted the pace of population change was then compared to now. Globally we are hurtling towards our crisis at a speed even now of around a 1% pa annual increase in population (it has quite recently been considerably higher, almost 2% in the 1970s) while in 1300 the increase was a stately one fifth of that. But the population declined in the crisis far faster than it had been going up (1% pa compared to 0.2% pa). If this pattern were followed in our crisis we might expect our population decline to be much more rapid than the ascent; if, like the Late Middle Ages it were five times more rapid we would be talking about a 5% pa decline, or a halving of population every 14 years. It is hard to conceive of a population catastrophe on that scale; it would exceed for example for the whole world the rate of population loss in the Soviet Union in the worst years of the Second World War.

Technology

A few paragraphs back I set aside the possibility that technology will allow us to avoid the rather stark population decline that the Medieval experience seems to suggest. Let us now back-track to examine that possibility. On the face of it, the absolute technological comparison between 1300 and now is simple; we have massively greater technological resources now than then. Moreover, not only do we have better technology, our technology is improving much faster, and perhaps most important, we are aware of the potential and importance of technology for solving our problems. This belief in technology, including technologies yet to be invented, perhaps lies behind a great deal of our current complacency.

There is some difficulty refuting the argument that technology will save us. It is of course fairly straightforward to refute some of the proposed simple extensions of existing or at least expected technologies. So for example, we are assured that transport will be electrified, and we will carry on using cars but electric ones, and that indeed such private transport systems will be extended over the whole planet. This can be refuted by pointing out that it will be beyond the likely capacity of any conceivable sustainable generation system to supply enough electricity, that the embodied energy in so many vehicles and so much highway is itself problematic, and that the production and disposal of sufficient batteries is likely to produce pollution on a hitherto unimaginable scale. The increased costs would unbalance the whole economy. And similar arguments can be had about other actually imagined specific technologies, such as limitless energy from renewable sources, huge agricultural yields from genetically modified crops, or schemes of geoengineering.

But no list of specific refutations can dispose of arguments that effectively cite entirely unknown future technologies that will save us—and claim that to believe they will perhaps not turn up is to be a pessimistic Luddite.

Yet these technologies, whatever they are, will still be bound by the basic laws of physics. An argument can be mounted based on the Laws of Thermodynamics that our technologies will, whatever they are, be overtaken by the increasingly impossible amounts of energy needed to prevent the ever-increasing amounts of entropy, or loosely disorder, that our activities create (Georgescu-Roegen 1971).

There are two counter-arguments. While there do seem to be quite low limits to the potential energy we can harness from renewable resources, there is one potential high technology energy source, hydrogen fusion, that might in practice provide an almost unlimited supply of energy. It's just that ever since the hugely expensive search to harness fusion began in the 1950s, the prospect of a real practical power source has always seemed at any one time to be at least 50 years away.

The other counter-argument to the energy/entropy argument is that it is far more difficult to prove that we are now actually close to this entropy limit (as we are close to catastrophic climate change), and in my view this has not been done.

Perhaps more important than this rather theoretical argument, it is surely simply irresponsible to assume that technologies will just turn up, and actually be applied to solve our problems. It's tempting to speculate that this is perhaps more like the fourteenth century than we might like to think: then we believed that the best answer to famine and plague was prayer to an almighty god who would set things right, while some began to think that it was not just prayer that was needed but less sin. Now many have a rather unexamined belief that technology and progress will solve our problems. Life has got better by and large year by year for the last 300 years, progress is normal, why should it stop? Both approaches are no more than unexamined faith, and if the fourteenth century has a lesson it is that such mistaken faith can be very persistent even when faced by very significant challenges. There is good reason for thinking that one factor behind the lack of actual action on climate change is a strong and persistent faith in this belief that something will turn up. We are perhaps more aware of our predicament than our medieval predecessors, but ironically faith, not in God but technology, may well blind us, especially in Europe, as much as ignorance blinded people in the Middle Ages.

Relations of production

The Middle Ages began with one system of production, feudalism, and ended with the beginnings of another, capitalism. A crucial point in the transition was the Crisis of the Late Middle Ages. The labour shortage in the countryside following the Black Death was pretty much the death knell for the feudal system of social relations in western Europe, in that peasants could refuse labour services and demand to be paid in money. The Peasants' Revolt in England in 1381, just one of many peasant revolts in this period in Europe, was essentially directed against attempts by landowners to restrict increases in wages. Also, spare land abandoned during the plague meant that peasants could *in extremis* leave their manor and colonise abandoned land. While remnants of serfdom still existed in parts of western Europe by 1500, by and large most peasants were at least legally free, able in theory to move about, and owing the local lord a rent but not much else in the way of services. Other factors beyond de-population following the plague years were also important, such as the wider use of money and changes in military technology (mainly the increasing use of firearms) that reduced the power of, and hence the necessity for, the feudal mounted land-holding knight.

Will climate change threaten capitalism as the dominant system of production as feudalism was challenged in the fourteenth century? This question has two sides. The more usual one is to ask whether policies designed to mitigate climate change and wider environmental destruction (e.g. an end to economic growth, sparser and more expensive energy, an end to the consumer society) are compatible with capitalist social relations—can we be green and capitalist (Klein 2015)? Less common is the question of whether the *consequences* of climate change, including, as in the Late Middle Ages, de-population, and dire consequences for some of the poorer parts of the world, especially Africa and the Middle East, which are important sources of raw materials for the rich world, will also affect the dominance of capitalism.

Moreover, to this complex mix we need to add two particular current trends, which have no analogue in the Middle Ages, and which some claim will have profound effects on capitalism. First of these is the idea that as so much of the value of so many things in our economy depends essentially not on the material they contain but upon the information they embody, and that because reproducing that information is virtually costless, the marginal cost of many items will reduce to zero. If this happens, it is argued, then capitalism becomes impossible, and markets will disappear as goods and services are distributed by sharing mechanisms (see Mason 2015).

Second, there is the march of robotisation. With the prospect of self-driving cars, autonomous drones delivering Amazon orders, expert systems and artificial intelligence and so on, it is predicted that not just manufacturing but also huge parts of the current service sector, now the major employer in richer economies, will simply not need human workers any more. What becomes of the workers in an economy that no longer needs them? How will sufficient demand be created to keep the robots in work if the people who are left are too poor to buy anything? Can a capitalist system exist if it has no economic place for most of the population?

These are all huge issues in themselves, but I'm going to confine myself here to asking whether the experience of the Later Middle Ages gives us any help. I think the most important point to make is that around 1300 many changes had already taken place that undermined feudalism—the spread of money, economic growth that took per capita incomes above subsistence for more than just the elite, more urban life, changes in military organisation and technology—which were then given a huge impetus by the depopulation of the mid fourteenth century and the economic and political opportunities which that offered to serfs. And capitalism, in the form of economic colonialism and the opportunities for 'primitive accumulation'—that is effectively the theft of resources from colonies—which it offered, received a further huge boost from roughly 1500 with the discovery of the Americas and technical advances in shipping.

Our current situation is not at all analogous. Capitalism is a robust and highly various form, ranging from neoliberalism in the UK and US to social welfare capitalism in France and Germany and to a form of state capitalism in China ('socialism with Chinese characteristics'), just to take some examples. It can mutate and change. The challenges it faces are many and various, and by no means all in the same directions. For example, reducing energy supplies suggests a greater demand for labour, especially for example in agriculture, while robotisation suggests a reduced demand. Basic things like food and shelter would take up a greater share of a smaller economy, yet these are the very things that don't have zero marginal costs; if I share my lunch there is less for me. A divided international world encouraging autarky may damage globalisation and the multi-nationals, but a more national form of capitalism with considerable state intervention, like Second World War Britain, is still capitalism. While in some parts of the world institutions like sound money and the effective legal defence of property, which are pre-requisites for capitalism, may be threatened, their roots in the rich world are deep and strong. Despite the depth of the 2008 financial crisis, the system has in fact carried on much as before: there's been a depression but not a revolution.

However, as we've seen above, de-population, especially in the poor countries is likely just as in the crisis of the Middle Ages. This led to the availability of surplus land, and in a world where land was necessary for most economic activity, strengthened the position of the poor. On balance the threat to capitalism is not like that to feudalism in 1300; capitalism may have to bend and change but it's hard to see in climate change a mortal threat to the system.

In concluding this section, we must also address gender. Europe in the Middle Ages was a deeply patriarchal society but unlike the economic relations of production, there is little evidence that this aspect was affected by the crisis of the Late Middle Ages. While the increasing opportunities as the population declined may have helped the position of women, including in particular increasing access to common lands, there appears to be no permanent change persisting past 1500.

What then is the implication of this for the position of women today as we face the crisis of climate change? For the parts of the world that are resolutely patriarchal, mainly peasant societies in the poor world, climate change and its attendant plagues and famines will heap great suffering on billions of women in particular, while the survivors, like those in the later 1300s may for a while face a slightly easier life, though no doubt still within the fetters of patriarchy. In the rich west, where patriarchy has been essentially undermined by women's participation outside the household in the labour market, there is nothing in the medieval experience to guide us.

Ideas 1300–1500

Drawing out the comparative effects of ecological crisis and climate change on ideas and cultures, upon what French historians call 'mentalités', we immediately have to narrow our focus and ambitions. First, it is way beyond this author's competence to discuss the effects of the coming storm upon all the contemporary world's mentalities. The connections of late medieval European thought are with what that system of thought became, that is broadly modern western thought, not with modern Buddhism for example. So I will restrict what I say to broadly western ways of thought, not discussing the Moslem, Hindu, Chinese and other non-Western world views. Some of these views may in fact be better adapted to our future world than western attitudes, with their emphasis on dominion over nature and individualism, but I shall not pursue that here.

Second, in many ways this is even more tentative than what has gone before, which is mainly physical, then economic and more tentatively political. Only the most doctrinaire historical materialists believe that how we think and what we think about are actually determined by our economic circumstances; I am one who believes that mentalities are seriously influenced by material events, perhaps even constrained by them, but also have important autonomous development of their own, probably impossible to predict.

Central pre-occupations

However, let us begin. How does how people thought in 1300, and how that developed, compare with our present starting point? The central point is the most obvious one, yet it can get lost when people begin talking about how ideas evolved after 1300. In 1300, for most people, high or low, for most of the time, whatever they were thinking about or doing was soaked in the dogmas and assumptions of Christianity, and Christianity manifested in the particular form of the universal Roman Catholic Church (Verger 1986). No part of life or thought escaped this, and even the new thinkers of the time largely reacted against it and operated within its framework. It was impossible to be an atheist, a liberal, a feminist, a socialist, a scientist or even an economist—these ideas did not exist.

This uniformity is not reflected in the relative plurality of modern day thought and attitudes; the very idea that an individual might arrive at a set of beliefs by their own thought and reason would have been regarded in 1300 as sinful, and potentially hugely damaging to society as a whole. Yet we in the richer world live in societies that are the product of centuries of hard-fought-for toleration, and that very possibility of diversity, and freedom of thought and speech is a basic part of our makeup. And while we might despair at the bland uniformity of much popular culture, and the ubiquity of materialistic consumerism, these diversities are real; current examples are the genuinely very different views of the world of US Trump supporters and 'liberals', or UK leavers and remainers.

The pre-occupation at the centre of the mentality of the Middle Ages was the issue of personal salvation; were you heading after death for eternal life in heaven, for a spell in purgatory before your fate was determined, or for eternal torment in hell. In the High Middle Ages your expectations were mediated by the Church. Most people were reasonably assured that if they did what the Church said, were not heretics, performed the necessary rituals, and, if they were rich, made appropriate contributions to charity, this central question would probably turn out alright, with a brief time in purgatory before joining the saints in heaven. They could sleep at night, and get on with their lives.

But in the Late Middle Ages the Church as an institution in many parts of Europe went into decline, its universality undermined by the 'captivity' of the Avignon Papacy (when the Popes fell under the control of the French crown), later split by the great Schism (when there were two Popes, which undermined the infallibility of both), and its presence on the ground weakened rather patchily by the depredations of plague, famine and war. The result was that people were still confronted with the issue of how to be assured of salvation, but were left much more to their own resources in order to obtain it. Mostly this led to an additional anguish to be laid over the very real problems of the here and now. For a very few, mainly intellectuals around people like Wycliffe and Huss, the seeds were planted of the idea of a new direct relationship with God, unmediated by the Church, which was to result in the sixteenth century with the Reformation. But this impending change meant little at this time to the mass of the people.

It is hard to point to a similar single focus for present-day mentality in Western Europe. Partly it is simply that after going through religious wars, the rise of doubt and the triumph of toleration, we live in a far more diverse ideological age. But there are some very common pre-occupations which we in the rich West nearly all share, which we could label as the essence of the mentality of modernity. We start from a negative contrasting point. Very few modern Western Europeans believe in a literal heaven and hell, and few are seriously worried about what will happen in their after-life. Their focus is on the here and now, and what as an individual they make of this life; it is the culmination of the development of individualism that began in the Late Middle Ages. As an aside this is very much a European view, not shared for example in the Hindu, Buddhist or Moslem worlds, or indeed by many in the Christian Americas, including the US. And life is conceived of as an individual progress, both social and individual; individuals are much pre-occupied with making the best of it, being in some terms successful. For many success is material, hence much concern about materialism and its consumerist variant, but for most it is still living a full family life, doing a fulfilling job and having a respectable and respected place (which involves an appropriate level of consumption) in a community.

This deep desire for a successful, fulfilled individual life taken in the round is nurtured by many of the institutions of our modern society, by education, by the idea of a career, by the importance placed on long term relationships, by aspiration to the middle class, by house ownership, by a long and contented retirement; we all have in short an aspiration to live a good life, even if we may

151

disagree about the details of what makes it up. It resides within the idea of progress, and the institutional basis making that idea a reality for all but the very rich is the modern European social state.

What will happen to this individual progress narrative of modernity when its material, economic and social foundations begin to fall away? The experience of the Late Middle Ages makes a suggestion, though no more than that, about the nature of the 'post-progress' condition. The essential point to take from the Late Middle Ages experience is that it took a very, very long time, say very roughly from around 1300 to the early twentieth century, for the central problem in most people's mind to switch from salvation after death to how to live a good life in this one. People cast around for new solutions to the old problem (the Reformation offered a different route to heaven, not an injunction not to worry about life after death) long before they began to think that maybe there wasn't a problem after all. That leads me to suggest that, despite a general quickening in the pace of social and intellectual change, we will mainly be still (in Western Europe at least) looking for individually progressive, modern, materially abundant and fulfilled personal lives in the here and now even when, as this century progresses, many of the fundamental props that support it will have withered away. While a few will become detached contented Buddhists, or resigned nature-worshipping pantheists, and that may be our ultimate destination, they will be only a small group of pioneers.

And this striving for the good life will cause us huge dissatisfaction and distress as it becomes less possible. It may well feed greater competitiveness and less concern for others, perhaps more cynicism. Some may begin to feel they have a special right to succeed, mimicking the extreme ideas of pre-destination of the early Reformation, and using that right to exploit others. It is not a happy prospect, but perhaps armed with the example of Calvinism we might actively seek to avoid it.

Past and future

We turn now away from the central pre-occupations of then and now to the second great theme of late Medieval mentalities. In the Late Middle Ages, the material crisis, the decline of hope and optimism, led people to dwell on the past rather than think about the future. Thus Gerson, a dominant theologian at the University of Paris in the 15th century, did not tend to look for inspiration to the great system of St Thomas Aquinas created just 150 years before, and was certainly not inspired by modern fads like William of Ockham's nominalism. His inspiration was the simplicity of the original fathers of the church (Verger 1986, 148). In knightly circles, as the utility of armed men on horseback declined throughout the period, the cult of knightly chivalry grows paradoxically ever stronger, harking back the exploits of the High or even Early Middle Ages (e.g. the cult of King Arthur in England), and creating an over-ripe culture (Huizinga 1924). The early humanists, like Petrarch, despised the recent reconciliation of ancient philosophy and faith and wished to get back to the purity of the ancients themselves.

We live now in an obsessively forward-facing culture, born of three hundred years of progress, and very different from the tradition-bound culture of 1300, where despite the progress since 1000, the world had nevertheless changed but slowly. Combined with our idea of individual lives as a progress we look forward to the time when will we banish disease and defeat the aging process, or even solve our ecological problems by somehow escaping the earth. As this all collapses, if the Late Middle Ages are a guide, there will be an increasing tendency to look back, and idealise a golden age, perhaps the period of massive growth, equality and growing opportunity that followed World War II for 30 years (Hobsbawm 1994, Part Two), or perhaps even further back to the Victorian world of grit, thrift, respect and respectability. There are already signs of this in the UK in the wave of nostalgia that has partly motivated Brexit. Such looking back is not necessarily undesirable—it is part of the illusion of progress that the past must always be worse than the present. But it is hugely important to pick the right past to look back to. The world of the egalitarian Golden Age is a far more humane and inspiring precedent than for example the dog-eat-dog capitalism of the mid nineteenth century, or the glory days of the British Empire. It will be important, like Plutarch in his revival of classical learning, to choose the right bit of history for inspiration.

We've already remarked on the intellectual, functional and political parallels between Scholasticism and neoliberal economics. Scholasticism largely died out as science advanced and the monopoly of the Catholic Church declined with the Reformation. Its relevance was challenged during the crisis of the Late Middle Ages, not least because more real problems of plague, war and famine, and the breakdown of social and political order, began to assert themselves. In parallel, the influence of laissez-faire and neoliberalism declined during the long world war from 1914 to 1945 and its aftermath in the Golden Age until around 1975, as people grappled with more immediate problems than an exclusively economic focus. We are seeing too some decline in neoliberalism now since the 2008 crash, but I suspect this may be transient. Nevertheless, the awful impending and longer-term crisis of the twenty first century may have one small silver lining too, the end of neoliberalism.

Conclusion

So what are we to conclude? How do the events of 1300–1500 help us understand how the world is likely to be in the next two centuries?

First, we note the obvious similarities. Two important features of Europe in 1300 and the world in 2000 are the same; most importantly over-population and broad ecological crisis but also climate change, albeit we face significant warming while medieval Europe faced just a little cooling. Second, we both faced this after periods of unprecedented economic growth and expansion. Third, both were preceded by periods of intellectual growth but also change; in the medieval period by the synthesis of Christian and classical thought, in our own times by the rise of science.

Second, there are massive differences. While overall both periods start with an ecological crisis of over-population and environmental degradation, in 1300 most people were desperately poor. In 2000 many countries are unbelievably rich by comparison with 1300, though for the majority of people who are poor, things are more similar. In some ways we understand our predicament quite well; in the 1300s there was little understanding of what had gone wrong. In particular, not only have we made huge technical progress compared to 1300, but we also understand the potential and significance of technical progress and research for solving our problems. But we share the intellectual complacency of the High Middle Ages in that we have rather too much faith in technology, just as faith in God led to inaction in the Middle Ages. So in these more human terms, in this matter of faith rather than knowledge, we are every bit as uncomprehending, but perhaps a lot more arrogant, as we speed towards our nemesis than were our fourteenth century predecessors. Indeed, as predominantly rural dwellers closer to realities, they probably coped better than we will with more than half of us now crammed into cities.

So which of the developments after 1300 are relevant for us. The first is simple: as in the 1300s, war, famine and pestilence are set to reduce world population after a period of unprecedented growth.

Next, while very broadly most of this decline will take place in the poor world, not the rich, if 1300 to 1500 is a guide the effects will be very patchy. The faith that as yet unknown technologies can solve our problems will prove very persistent, just like faith in prayer in the Late Middle Ages. We will struggle with the fact that our individual lives no longer show progress, and will tend to look for models for how to behave much more to the past than we do now.

Finally, thinking about the Late Middle Ages presents one obvious fact not so far mentioned here. Terrible as those times were, life went on. People lived, loved, had fun, hoped, suffered and despaired, planned, plotted and intrigued, fought each other for often rather trivial reasons and died, often rather oblivious to the social and economic currents around them. It is the writers of fiction from the period, Boccacio with his *Decameron*, or Chaucer with his tales of the Canterbury pilgrims that best exemplify this. They are not full of gloom, they are full of life. Both show life in all its diversity, and the mentions of war, plague, economic ruin and so on, all the larger events that have pre-occupied this piece, are few. Nearer our own time Tolstoy made the same point about the awful events of 1812 in Russia in *War and Peace* (Tolstoy 1849, 1116):

> With half of Russia in enemy hands, and the inhabitants of Moscow fleeing to distant provinces, with one levy after another being raised for the defence of the Fatherland, we, who were not living in those times, cannot help imagining that all Russians, great and small, were solely engaged in immolating themselves, in trying to save their country or in weeping over its downfall. All the stories and descriptions of those years, without exception, tell of nothing but the self-sacrifice, the patriotic devotion, the despair, the anguish and the heroism of the Russian people. Actually, it was not at all like that. It appears so to us because we see only the general historic interest of the period, and not all the minor personal interests that men of that day had. Yet, in reality, private interests of the immediate present are always so much more important than the wider issues that they prevent the wider issues which concern the public as a whole from ever being felt—from being noticed at all, indeed. The majority of the people of that time paid no attention to the broad trend of the nation's affairs, and were only influenced by their private concerns.

Life will go on, at least for a while.

Chapter 8

FACING UP TO ECOLOGICAL CRISIS: A PSYCHOSOCIAL PERSPECTIVE FROM CLIMATE PSYCHOLOGY

Nadine Andrews and Paul Hoggett

Introduction to climate psychology

Facing the facts of climate change and ecological crisis involves encountering powerful feelings such as loss, guilt, anxiety, shame and despair that can be difficult to bear. How we deal with these feelings shapes how we respond to the crisis and is critical in determining whether these responses are ultimately adaptive or maladaptive. Adaptive responses support psychological adjustment to the emerging new realities and stimulate appropriate and proportional action (Crompton and Kasser 2009). Maladaptive responses work against this in some way.

In recent years the psychological dimensions of climate change have received greater attention in climate science, policy and in the media, with growing acceptance and appreciation of the contribution that these perspectives can make in enriching our understanding of both the causes and the impacts of climate change. The field has advanced from arguing that psychology has an important contribution to make (American Psychological Association 2009), complaining that the contributions and insights are not widely accepted or applied (Swim et al. 2011), restating the argument in mainstream scientific publications (Clayton et al. 2015), to a position of greater influence over the inclusion of psychology in Intergovernmental Panel on Climate Change (IPCC) reports. For example, the American Psychological Association has achieved IPCC observer organisation status and some leading psychologists have been selected as authors for the Special Report on 1.5 degrees and for

157

the main sixth assessment report. The significance of this development should not be understated, but nonetheless the role that psychology is currently playing still barely scratches the surface of its potential. There is much more that psychology—though not necessarily mainstream psychology—can offer. This claim brings us to the purpose of this chapter: making the case for the contribution of what we term 'climate psychology'. This approach has emerged in recent years occupying a space left under-explored by mainstream psychology and climate science: the role of the non-rational and unconscious in human cognition and decision-making. These aspects are not entirely ignored by the mainstream of course; constructs like 'denial' are popularly used to explain lack of engagement and adaptive action, but a deeper more nuanced discussion is generally absent from the literature. For example, denial (in the context we mean here) features just once in the Working Group II[1] contribution to IPCC's Fifth Assessment Report: 'If a perceived high risk is combined with a perceived low adaptive capacity, the response is fatalism, denial, and wishful thinking' (IPCC 2014b, 204). Its single appearance in the Working Group III[2] report is slightly more elaborated: 'There is evidence of cognitive dissonance and strategic behaviour in both mitigation and adaptation. Denial mechanisms that overrate the costs of changing lifestyles, blame others, and that cast doubt on the effectiveness of individual action or the soundness of scientific knowledge are well documented, as is the concerted effort by opponents of climate action to seed and amplify those doubts' (IPCC 2014a, 300). Neither of these entries explains the psycho-social processes involved, nor how to address them.

The reason why mainstream psychology struggles to account for the complexities of human behaviour is largely due to its methodological conventions. Studies often split off individual behaviour from its social context and focus on a single factor or a small number of factors. It is largely quantitative, relying on self-report survey-based methods or controlled laboratory studies which usually use students as subjects. Climate psychology on the other hand uses qualitative methods and proceeds from the premise that when it comes to understanding humans it is precisely what can't be counted that really counts. It draws upon a variety of sources that have been neglected by mainstream psychology including psychoanalysis, Jungian psychology, ecopsychology, chaos theory, continental philosophy, ecolinguistics and social theory. From such sources we glimpse some of the complexity and mystery of the human. The raw passions that often dominate our thoughts and behaviours; the internal

---

1 IPCC Working Group II report is on impacts, adaptation and vulnerability.
2 IPCC Working Group III report is on mitigation.

conflicts and competing voices that characterise our internal lives and give colour to our different senses of self; the effect of systems of domination on the way we think and feel about ourselves. Viewed from this perspective it is possible to see how our attempts to defend ourselves against the feelings aroused by worsening climate change are mediated by deep-seated assumptions about ourselves and society. For example, a powerful sense of entitlement may help us to shrug off guilt and shame, or a touching faith in progress can mitigate anxiety and induce complacency. Typically we will feel torn between different impulses, to face and avoid reality, between regret and the desire to make amends versus cynicism and hopelessness, between what is convenient for us and what is necessary for the common good.

Climate psychology attempts to offer a psycho-social perspective, one that can illuminate the complex two-way interactions between the personal and the political, the psychological and the social. It is concerned with understanding how our collective paralysis plays out in both our individual lives and in our culture.

The contribution of climate psychology

Climate psychology seeks to further our understanding of:

- collectively organised feelings such as loss, despair, panic and guilt evoked in individuals, communities, nations and regions by climate change and environmental destruction;
- defences and other strategies used to cope with the psychological threat brought on by climate change and ecological crisis, such as denial and rationalisation, that help us avoid facing difficult feelings, and how such threat responses manifest in the lives of individuals, families, groups and institutions;
- cultural worldviews and practices (e.g. faith in progress; sense of privilege and entitlement; materialism and consumerism; short-termism) that are ecologically damaging and inhibit effective change;
- conflicts, dilemmas and paradoxes that individuals and groups face in negotiating change with family, friends, neighbours and colleagues;
- psychological resources that support change: resilience, courage, radical hope, and new forms of imagination.

Climate psychology aims to:

- promote creative approaches to engaging with climate change and thinking in a realistic way about something whose implications are unthinkable;

- contribute to changes at the personal, community, cultural and political levels;
- support activists, scientists and policy makers seeking to bring about change;
- build psychological resilience to the destructive impacts of climate change, both those already being experienced and those that are anticipated, so that people can prepare and adapt as best they can, in ways that go beyond individual self-protection to serve the greater whole.

We now discuss each of these areas of understanding in turn, beginning with collectively organised feelings.

Collectively organised feelings

When we think of emotions we typically think of them as 'belonging' to the individual somehow. But why should emotion be individualised in this way? After all, we don't see language and meaning as private to the individual. A psycho-social perspective insists that emotion is as much a public as a private phenomenon, that powerful collective feelings can be provoked by social change and can also contribute to social change (Hoggett 2009). In the last couple of decades sociologists have become increasingly interested in the collective organisation of feelings (Goodwin, Jasper and Poletta 2001; Gould 2009) and several ways of thinking about this have emerged. First there is the idea that a particular constellation of feelings may characterise a historical era perhaps spanning decades. Following the cultural historian Raymond Williams this is often referred to as a 'structure of feeling' (Williams 1961). An example would be the *ressentiment* (a cocktail of grievance, envy and anger) that characterised much of German society in the 1920s and 1930s after the humiliation of the Treaty of Versailles which ended the First World War (Coser 1961). Secondly there is the idea of 'abiding affects' (Jasper 1998), feelings which endure over shorter timescales and may be specific to particular social groups such as social classes, minorities, and so on. A good example is provided by Debbie Gould in her (2009) study of the gay and lesbian community's response to the AIDs crisis in the USA, the powerful role of shame in demobilising protest and of 'gay pride' in sustaining and advancing it. Last but not least, contagious collective feelings such as panic may find expression in society in sudden and fleeting ways. The origins of crowd psychology lay in the study of such phenomena (Le Bon 1896), feelings which today are increasingly amplified by social media.

Emotion and affect

With regard to climate change, much has already been written on the powerful feelings evoked by it including grief (Randall 2009; Head 2016), melancholia (Lertzman 2015), fear and terror (Doppelt 2016). Of course many of these feelings are only aroused once someone becomes aware of climate change, because it is in the nature of an emotion that it has a conscious object (climate change) and meanings (risk, loss) attached to it. But there is another kind of feeling, commonly referred to as an 'affect', which operates at a much less conscious level. Anxiety (the core ingredient of stress) is a classic example: typically when we are anxious our feeling flits from one thing to another, it has no secure object to attach itself to. It is also very visceral, felt primarily through the body rather than through cognition. This distinction is important because it enables us to understand how people, such as those in denial, may not be 'climate aware' and yet nevertheless affected by powerful feelings provoked by climate change.

Containing our feelings

The distinction between emotion and affect is helpful for another reason. Generally speaking the more an emotion can be made conscious and therefore subject to reflection the more it can be contained. 'Containment' refers to the extent to which an experience can be digested and worked through or, put another way, the extent to which a feeling can be transformed into an emotion as opposed to remaining as an affect. The psychoanalyst Wilfred Bion (1962) likened the mind to a system for digesting experience. If an experience can be contained, even a difficult one, it will provide food for thought (and therefore for growth and development). But if it can't be contained then, like indigestible food, it gets stuck in the system or has to be expelled. So if we can't contain feelings of sadness or terror they will either get stuck in our system (paralysing despair) or we will get rid of them by projecting them into others (we find a substitute object, the Other, for our fear) or through blind forms of action (because action can itself be a way of getting rid of uncontainable feelings). Simply put, are we able to use our feelings or do they use us? Are we able to regulate them and use them creatively or are we going to be at their mercy? Crucially, this brings us on to the theme of the next section: the way we use coping and defence mechanisms when faced with the difficult-to-contain feelings aroused by climate change.

Coping with psychological threat

Ecological crisis: when the environment of a species or population changes in a way that destabilises its continued survival.

— Isildar (2012)

161

We are without doubt in a situation of crisis. Destabilisation of planetary cycles and processes that regulate Earth's life support systems poses profound psychological threat: for example, to our sense of safety, and to the integrity and stability of self-identity, for it disrupts who we may think we are—as human beings separate and superior to other species with mastery over nature and control over our own lives. In challenging the morality of our destructive behaviours, our complicity and inadequate responses, it also poses a threat to self-esteem, denting our confidence in our own worth and our abilities. Ecological crisis also threatens our life plans, ideas of progress and internalised expectations of the future. It is abundantly clear that how we have lived until now cannot remain the same for much longer—many aspects of our lives will have to change fundamentally in order to mitigate and/or adapt to the impacts of destabilised natural systems (Crompton and Kasser 2009; Lertzman 2015; Weintrobe 2013). Ultimately ecological crisis threatens our continued existence as a species.

When encountering a perceived threat, the disequilibrium caused creates stress, which is both physiological and psychological. The human tendency is to attempt to alleviate stress and decrease negative emotions through defence mechanisms and coping strategies in order to return to baseline functioning as soon as possible. In psychology literature the terms 'defence' and 'coping' are often used interchangeably. Whilst there is an argument for a distinction between them, namely that defence mechanisms are unconscious and unintentional and coping strategies are conscious and intentional (Cramer 1998), in lived experience threat responses are likely to involve both conscious and unconscious dimensions. The processes involved are also dynamic: there is possibility for movement of information back and forth between unconscious and conscious parts of the mind through processes of suppression and awareness.

Types of response

There are different ways of categorising defence and coping responses: for example, by classifying into avoidant and approach types. Avoidant coping is a defensive form of regulation, involving denial, distortion, disengagement, and—as we introduced at the start of this chapter—suppression of negative emotions. Approach coping has three predominant forms: active coping, which is direct action to deal with stressful situations; acceptance, which is cognitive and emotional acknowledgement of stressful realities; and cognitive reinterpretation, which involves learning or positive reframing (Weinstein, Brown and Ryan 2009). Approach coping is generally considered adaptive because effort is directed towards containing the anxiety-evoking situation and overcoming the stress associated with it, whereas avoidant coping, whilst it may relieve stress in the short-term, if prolonged or situationally inappropriate is likely to become maladaptive and serve pathological ends. These ends manifest at the individual level: avoidant coping has been found to be associated with poorer health (Weinstein and Ryan 2011). But there are also ecological implications: coping responses either support psychological adjustment to the emerging new realities and stimulate appropriate and proportional action, or they work against this in some way, and serve to protect us from having to make radical changes or take significant action.

We can also make a distinction between proactive and reactive coping. Proactive coping, also known as anticipatory adaptation or psychological preparedness, is undertaken in anticipation of an event, whereas reactive coping occurs after. The two types merge when responses are made to an event in order to both diminish its impact and prevent its re-occurrence (American Psychological Association 2009). Coping responses can be cognitive, affective or behavioural—or a mix of these.

Ecologically adaptive or maladaptive

The literature on the psychological dimensions of climate change and environmental behaviour identifies various defences and coping responses, which can be categorised as follows.

Ecologically maladaptive responses:

- Denial or disavowal of ecological crisis (e.g. rejecting, deflecting, ignoring)
- Distortion of facts (e.g. reducing size of threat, putting threat into the future)
- Shifting responsibility (e.g. blame-shifting, denial of guilt, splitting, projection)
- Avoidance of difficult emotions (e.g. suppression, rationalisation, escapism, numbing, pleasure-seeking)
- Diversionary activity (e.g. minor behaviour change or displaced commitment)
- Non-action (e.g. resignation, passivity, lazy catastrophism)
- Self-deception (e.g. wishful/magical thinking, unrealistic optimism)
- Active catastrophism and self-destructive acts
- Self-enhancement values orientation (e.g. materialistic behaviour to enhance self-esteem, or self-protection to enhance sense of security and being in control).

Ecologically adaptive responses:

- Seeking information, engagement with facts about ecological crisis
- Engaging with and regulating associated feelings (e.g. through mindfulness)
- Compassion, self-transcendence values orientation (care for human and non-human others)
- Connecting with nature
- Considered reflection on death, impermanence, human frailty and limitations
- Collaborative problem-solving.

The psycho-social nature of threat responses

Threat responses are not isolated psychological processes: they are psycho-social phenomenon, culturally sanctioned and maintained by social norms and structures (Randall 2013; Crompton and Kasser 2009). As Lertzman (2015, 3) explains, 'we produce, share and co-construct our unconscious negotiations of highly charged issues through our conversations, stories, advertising, intimate dialogues and public media discourses'. Threat responses interact with other psycho-social factors in complex ways to influence cognition and behaviour (Andrews 2017). Understanding these processes and their dynamics and effects is critical for designing interventions to subvert maladaptive responses to ecological crisis, at individual and group levels. Becoming aware of maladaptive responses as they arise within us offers the possibility of choosing a different response. We discuss this further in the section on 'Psychological resources'.

Cultural worldviews and practices

Climate psychology is interested in the beliefs, values, assumptions and presuppositions that are held at the collective level of human culture (cultural worldviews), and in the social practices that proceed from them. A cultural worldview serves to convey a sense of living in a world of order and meaning, giving purpose and a sense of meaning to life (Solomon et al 2004). The worldview that is dominant in a society pervades all aspects of life, and constitutes the broader contextual forces influencing each individual's values, identities, threat responses and conceptual systems. But the influence of a worldview often goes unnoticed by the individual, its assumptions unchallenged (Griffin 1995). This is because with repeated and ubiquitous exposure to its discourse, the worldview becomes internalised and is no longer obvious. It is taken for granted as something natural, rather than as something that we created and can therefore change. Climate psychology is concerned with making these assumptions visible so that they can be examined.

Ecologically damaging beliefs and assumptions

In western industrialised cultures there are a number of beliefs and assumptions that we contend drive ecologically damaging practices and inhibit effective change:

- The project of modernity aims to achieve human self-determination and freedom through a struggle against the limits imposed by nature.
- Nature is a wild malevolent force that threatens civilised life.
- Humans are entitled to dominate and exploit the natural world for our own ends—it exists for our use.
- Humans are different from and superior to nature, and we can free ourselves from the limits of nature through our ingenuity.
- We can control nature and harness its forces through technoscience.
- Advancement of technoscience, industrialisation and economic development (achieved through a free market economy) are vital for human progress i.e. for improving the human condition.
- Accumulation of material wealth and consumer goods is the primary measure of success, status and worth.
- Problems created by the above (e.g. human poverty and inequality, overwhelm of biosphere cycles and processes) can be solved by doing more of the same (e.g. techno-fixing our way out of climate crisis).

Disconnection and disregulation

As Schwarz's model of self-regulation of individual health explains, disconnection in feedback makes it more difficult for a system to self-regulate (Shapiro and Schwartz 1999):

Disattention → disconnection → disregulation → disorder → disease

Through repeatedly not attending to an emotional, physical body or interpersonal cue, the individual becomes habituated to the stimulus so it is no longer accessible to awareness, and disconnection in feedback occurs. The model can be applied to the human relationship with the natural world. Disconnection from the natural world and denial of natural limits reduces our ability to regulate human activity within planetary boundaries: the Earth's feedback signals are ignored, misinterpreted or simply not noticed. Symbiosis becomes overexploitation, ingenuity turns into hubris, and human interests always win out to the detriment of other species.

Wild and malevolent nature

The belief that humans are different from nature is closely associated to a conceptualisation of nature as wild and uncontrollable, a malevolent force that needs to be fought against (Rust 2008). This perception manifests in language, for example with metaphors that equate nature with evil, projecting human vices onto the natural world. As the moral philosopher Mary Midgley (2003) explains, such projection not only absolves a person or society from destroying nature but also fulfils another psychological purpose: by killing the personification they have killed the vice, 'they are symbolically destroying their own wildness' (166). A perception of malevolence is also conveyed in the phrase 'fighting climate change' where the climate system is personified as an active wilful enemy with intent to harm (Andrews 2016). It is to such a malevolent entity that the title of Lovelock's book *The Revenge of Gaia* speaks (Macy 1993). Our culture tames wildness and instils order through domestication of plants, animals and people (Totton 2011; Midgley 2003). These efforts give the impression of control over nature, and by extension the illusion that we can transcend its limits.

Mortality threat, control and self-protection

The ultimate natural limit that we may seek to transcend is death. We observed earlier that climate and ecological crisis poses existential threat. Unless the sense of self as part of nature is strong, existential threat tends to motivate people to enhance their self-esteem by prioritising self-enhancement values and pursuing extrinsic goals of material wealth and success (Fritsche and Häfner 2012; Sheldon and Kasser 2008), thus increasing consumerism and consumption and driving further depletion of nature's resources in a hellish self-reinforcing feedback loop. But climate change not only confronts us with our reality as mortal beings, it also reminds us that we are at the mercy of much more powerful forces. In this way it punctures hubristic illusions of control, destabilising our sense of who we think we are as human beings separate from and superior to nature, and capable of controlling nature through our inventions in technoscience. The facts of ecological crisis are so resisted because they confront us with the reality that this deeply entrenched and long-standing cultural worldview *is no longer tenable*. As Naomi Klein (2014) says, climate change is a crisis of civilisation: a crisis of story.

One response to this crisis of story that we can see becoming stronger at both group and nation state level is self-protection. Self-protection values and isolationist tendencies have erupted to the surface creating political chaos, for example with the presidency of Donald Trump in the USA and the vote by Britain to leave the European Union. Self-protection is anxiety-based, and therefore always has a paranoid tinge, it is pursued to cope with situations of uncertainty (Schwarz et al. 2012) and it can easily lead to physical violence.

Entitlement, resentment and social conflict

The rise of technoscience can be dated back to the mid-1800s at the end of the Industrial Revolution (White 1967). This union between science and technology enabled exploitation of nature on an unprecedented scale, way beyond that of the Scientific Revolution in the preceding centuries. A sense of entitlement to exploit nature helps shrug off any guilt that may be felt about the ecological destruction and disruption caused by overexploitation and over-consumption of nature's resources.

Gregory Bateson (1982) observes that the Industrial Revolution came with an 'enormous increase of scientific arrogance' (311). Arrogance, superiority, exploitativeness and entitlement are dimensions of narcissism (Weintrobe 2004; Frantz et al. 2005), which can apply to societies and organisations as much as to individuals. With the rise of consumer culture, narcissism has become more prevalent in society (Lasch 1979; Kanner and Gomes 1995). Narcissism has been found to be an inhibitor of nature connectedness and a major barrier to resolving environmental problems (Frantz et al. 2005).

There are other ways in which a sense of entitlement can inhibit effective climate action. The loss of something to which we feel entitled, such as certain levels of material wealth and living standards, can trigger perceptions of injustice and feelings of resentment towards 'liberals' who urge restraint and living within our ecological means. Such grievances increasingly fuel support for Trump and other populist climate deniers.

A much greater danger is now appearing on the horizon. It is highly likely that we are about to enter an era of 'ecological austerity' in which the effects of climate change and other ecological disasters lead to an inexorable rise in food prices, the mass displacement of peoples and the end of economic growth as conventionally defined. No longer 'finger-wagging liberals' but the earth itself will force restraint upon the entitled citizens of the West. The risk is that this fuels a new era of *ressentiment*, akin to that which underlay the rise of the far right in Europe in the inter-war period. When governments fail adequately to protect against or respond to climate change impacts, this may lead to loss of faith and trust in governments and other civil institutions, resulting in backlash. Perceptions of inequalities or disparities in the impacts of climate change may trigger social unrest, intergroup conflict and violence (Doherty and Clayton 2011).

Trapped in ideology

The idea that technoscience can solve problems created by technoscience has been termed an 'ideological pathology' and a progress trap (Wright 2004). This trap also applies to capitalism, which sees the answer to climate crisis as even more capitalism, turning nature and even carbon emissions into commodities. Organisational studies scholars Christopher Wright and Daniel Nyberg (2015) call this process 'creative self-destruction' whereby global capitalism devises ever more ingenious ways to exploit and consume the earth's resources and life support systems. As Jared Diamond (2005) discovered in his work on the collapse of civilisations, when faced with a crisis they do not understand, civilisations tend to reinforce the very routines that put them into that crisis, through force of habit rather than by reason. In times of disruptive change, these routines become a trap that becomes deeper and harder to get out of with the short-termist tendencies of our political and economic systems.

The (in)coherence of worldviews

Belief systems are not necessarily coherent and can contain contradictory views. For example, at the same time as believing that human ingenuity will triumph through technoscience, with climate change denial there is also a rejection of the authority of science (Hamilton 2010). Whilst the belief in human separation from nature underpins technoscience, it is science that also shows humans to be firmly connected with the living world, for example with Darwin's theory of evolution, and more recently with comparison of the genomes of humans with other species and the discovery that the human body comprises ten times as many microbial cells as human cells. However, the beliefs and assumptions we have discussed above predominate, and together they form a particularly environmentally toxic worldview.

Conflicts, dilemmas and paradoxes

All of us in the West are thoroughly implicated in the climate change crisis. In modern, technologically saturated societies we are completely dependent on the systems which bring energy to our kitchens and food to our shops, transport us to work, dispose of our waste, and so on. Under neoliberal capitalism, markets infiltrate everything, getting right underneath our skin, converting more and more aspects of life into calculation and exchange. And with globalisation the middle classes find their friends, colleagues and loved ones increasingly spread across the world. To 'drop out' of this system is virtually impossible and the vast majority of us, even those very 'climate aware', have to act for the best when faced with a myriad daily choices regarding how we travel, eat, communicate, socialise, love, relax and work. We also live in an increasingly plural society in which thousands of different voices clamour for our attention, drawing upon our different identities. Annie, an older, Welsh nationalist, conservationist, rugby-loving, Presbyterian, middle class, socialist, mother of three will be pulled this way and that by the many moral claims she feels are made upon her by all the different communities that she belongs to. According to the political philosopher Bonnie Honig, today we in the West live in 'dilemmatic space' in which there is often no obvious right thing to do (Honig 1996).

Take the dilemma of 'love miles', for example. Say Annie's only daughter now lives in Australia and is the mother of Annie's only two grandchildren. No matter how 'climate aware' Annie might be, no matter how conscious of the carbon emissions incurred by flying to the other side of the world, the 'pull' for her to go and see her family will be enormous. Whichever choice she makes she will be haunted by the loss and guilt incurred by the 'path not chosen', something the moral philosopher Bernard Williams referred to as 'the remainder' (Williams 1981).

A psycho-social perspective

Drawing attention to the conflicts and dilemmas that we all face as ethical beings in modern societies also enables us to understand something about what it means to be human. Rationalist models of the person that still dominate most academic psychology tend to adhere to a unitary view of the human subject. In contrast, a psycho-social perspective understands the human subject as always in tension, between conflicting needs, and between competing moral claims. It follows that, far from there being a unitary self which 'acts' and 'chooses', each of us is torn between a number of competing selves and identities, for example, between a complacent and entitled self on the one hand and a concerned and loving self on the other. And the problem for us all is that it is not at all easy no know which part of us is being complacent and entitled, and when. For example, if Annie decides not to go to Australia, is this because her concerned and loving self has triumphed or is there something cold and unfeeling about her decision? And by not going, is it possible that her entitled self has actually triumphed, one which views her as an exceptionally moral person, perhaps saintly in her forbearance of ordinary pleasures?

To reiterate the point made by Bernard Williams, in a complex, plural society there is often no obviously right thing to do; all we can do is 'act of the best' as he puts it, knowing that we might be wrong but not allowing knowledge of our fallibility to undermine our capacity to act in what we feel is the right and ethical way. Understanding the person in all their complexity provides us with a sound platform for engaging with others in conversations about climate and ecological crisis.

Knowing that we are no different from the other we seek to influence and change puts us in a more humble position from which to engage, and this in turn lessens the likelihood that the other will become defensive and increases the likelihood that we might both engage in a richer conversation about our predicament.

Psychological resources

Climate change constitutes a profound psychological challenge to all of us. It does this in two ways. First, awareness of climate change itself, its already occurring impact and the severe threat it poses to our collective future, can and often does feel quite overwhelming, and this is as true for the ordinary person in the street as it is for a climate scientist (Hoggett and Randall 2018). Secondly, the actual impacts of climate change, including drought, wildfires, floods and storms and rising sea levels, often challenge both our physical and psychological well-being. Climate change threatens us with experiences that are difficult to contain and unless we build up our resilience these will be potentially traumatic.

Resilience

Much of the research on resilience derives from studies of families living in extreme poverty and social deprivation. These are sometimes referred to as 'traumatogenic environments' because the multiple stressors (poverty, drugs and alcohol, violence, social breakdown) combine to create an environment which constantly overwhelms peoples' psychological resources. This research has now become incorporated into fields including development studies, social policy, genetics, health, education and disaster studies (Mohaupt 2008; Luthar et al. 2000).

One of the most significant concerns is that there is currently no single definition of resilience. One commonly used definition (Luthar et al. 2000: 10) is that resilience is a 'dynamic process encompassing positive adaptation within the context of significant adversity'. There must have been exposure to significant threat or severe adversity and the achievement of positive adaptation, despite major assaults on the developmental processes. But debate, for instance, surrounds the idea of what constitutes 'positive adaptation' and therefore whether in some circumstances resilience might include aggressive, deviant and what some might see as 'anti-social' behaviour.

Mentalisation

The literature on trauma, stress and coping highlights the personal resources crucial to human adaptation. A vast and varied body of literature on child development, adult mental health and personal self-help is dedicated to cultivating psychological resources to build resilience and support personal change. Psychotherapies, counselling, mindfulness practice—there are many ways we can intervene to help ourselves and others. Much of this knowledge can be usefully drawn upon to help us engage in a productive way with the conflicts, dilemmas, threats and impacts of ecological crisis.

Core to these approaches is the capacity for 'mentalisation'—bringing what was operating in the dark into the light where it is visible and can be examined, reflected upon, contained and regulated (Fonagy et al. 2002). In relation to trauma this involves the ability to get a distance from the disturbing experience without disassociating from it, and this can be enhanced by being able to deploy perspective: seeing things in a different way and from another point of view. The role of a trusted other—an individual, family or group—can be crucial in this process, and this shows us is that resilience is as much a relational phenomenon as it is something 'belonging' to the individual.

Transformational resilience

Thinking about resilience as a property of a social system takes us to the concept of 'transformational resilience' (Doppelt 2016), which can be applied to both individuals and communities in terms of their capacity to adapt to the impacts of climate change. Here, adversity is used as a catalyst for finding new meaning and direction in life, and changes are made that increase individual and community well-being above previous levels (Doppelt 2016). This is a particularly powerful way of thinking about how to deal with traumatic experiences and is already being used in northern Californian communities affected by wildfires.

Radical hope

Despair and powerlessness go together. If people feel that they have some power in a situation they are less likely to feel despair. Ultimately despair comes from the feeling that you simply don't have the resources (the determination, the courage, the knowledge, the skill, the support from others) to do anything about the destruction all around you. Despair is a form of inner defeat. But strangely enough, hopelessness does not necessarily lead to despair—it can be liberating.

The social critic Christopher Lasch once said that the worst is what the hopeful are always prepared for. The hope that comes from being able to face the worst is an enduring hope because it is not built upon a scaffolding of illusion and wishful thinking. It is defiant and courageous and it refuses to capitulate to what might seem like hopeless odds. 'Active hope is something we do rather than have', so say Joanna Macy and Chris Johnstone in their book *Active Hope* (2012), and this is exactly what the Italian Marxist Antonio Gramsci meant by 'optimism of the will', that resilient 'keeping on going on'. It is also the kind of hope canvassed by John Foster in the concluding chapter of this book, where he links its possibility for us in present conditions to a renewed capacity to recognise our situation as tragic—a recognition which, as Nietzsche knew, is actually the reverse of fatalism.

We can't know what the future holds. When people begin to emerge from the ruins of a life, they can see painfully what has been lost, but ahead is only an uncharted sea. But move on they do, and some kind of elemental confidence about life is slowly restored. In his book *Radical Hope: Ethics in the Face of Cultural Devastation* Jonathan Lear depicts a whole community, the Crow Indian in late nineteenth-century North America, coming to terms with the tragic destruction of a way of life and imagining a new future for themselves. Lear calls this 'radical hope': hope is directed towards a 'future goodness of the world that transcends the current ability to understand what it is' (Lear, 2006: 103). Radical hope is not just about determination and courage; it is also about love and a re-finding of all that is benign in the world.

Conclusion

Climate psychology aims to gain insight into the conundrum of why ecologically maladaptive behaviours persist in the face of compelling evidence that these behaviours are dangerous to life. With a focus on the role of the non-rational and unconscious in human cognition and decision-making, and an appreciation of the fundamental entanglement of psychological processes with social and contextual forces, climate psychology brings perspectives often overlooked by mainstream psychology. In this chapter we have introduced some core ideas: the role of emotions and how feelings are both personally felt and collectively organised; that ecological and climate crisis poses psychological threat and ways of coping can be adaptive or maladaptive; that there are cultural beliefs and assumptions that drive ecologically damaging practices and inhibit effective change; and that conflicts, dilemmas and paradoxes are inherent to being human. Finally we discussed psychological resources that can help build resilience and support positive change at individual, community and societal levels.

We are embedded in a cultural worldview that is fundamentally pathological, and it affects all of us. Acknowledging the conflicts and dilemmas that we all face as ethical beings in modern societies enables us to understand something about what it means to be human. None of us is completely right or completely wrong in the way we are responding, and the humility such a realisation engenders may help us relate to one another with more compassion and forgiveness.

Chapter 9

WHERE CAN WE FIND HOPE?

John Foster

The challenge

Suppose you are committed to struggle for a hugely important goal, but when you pause, step back and honestly assess your prospects, you recognise that the odds are overwhelmingly against your succeeding. What is then the realistic thing to do? To admit defeat and settle for the best you can get, however far short of your intended goal that may fall?—or to go on fighting whatever the odds?

Note that I am not asking what it would be brave or noble to do in such circumstances. I'm not asking what an outnumbered Anglo-Saxon warrior might do—'*Mod sceal þe mare, þe ure mægen lytlað*', as the man says in *The Battle of Maldon* (Gordon ed., 1937: 61): 'Spirit the greater as our strength ebbs'—nor, coming closer to home, am I asking what a hero in some Hollywooden epic would do—but rather, what it would be *realistic* to do. However inspirational the models we draw on, we must always remain realists—because, heroic fantasy aside, the world always remains stubbornly what it is, and ignoring how it really is gets you, in anything but the quite short term, precisely nowhere. But how *is* it, really? *Is* the world such that there is no bucking the empirically-calculated odds, the odds based firmly on experience of what has happened hitherto, when they are massively stacked against you? Or can sufficiently determined human action transform the limits of the possible? Can just going on fighting against huge odds *change* the odds?

These, in the context of this book, are clearly not mere hypothetical questions. The most vital—literally vital—political and social goal today is to prevent anthropogenic climate change from running away with us to the point where a human and ecological catastrophe ensues. And the odds against our succeeding begin to look not just enormous, but decisive. Yet it is very hard not to think that somehow, against these odds, one must find some realistic way to go on fighting—which means, going on hoping.

It is especially hard for me not to think that, for a personal reason which I think has wider-than-personal implications. The reasons for environmental activism, certainly in the kind of privileged first-world life which I guess most of my readers lead, have always depended on intellectual vision; the scientific modelling, the observable incremental deterioration needing to be *interpreted* into motivation. We could contrast here the spurs to socialist activism in the early twentieth century. The wide range of people who brought the Labour Party into being felt and saw and smelt the effects of exploitation, poverty and sub-standard housing, often because they themselves had to live under these conditions. They had explanatory theories, but these were instrumental to doing something about what had already motivated them by presenting itself as humanly intolerable. And it is fair to say that almost nothing about oncoming global warming yet presents itself, here in Britain, in our ordinary everyday lives, as humanly intolerable.

Almost nothing. But at the time of writing this chapter, I had just become a grandparent for the first time. And when that little lad reaches the age which I am now, it will be the later 2080s, and the world is likely by then to be experiencing all the disastrous consequences of a 4°C rise in global atmospheric temperature—as noted in the Introduction, the Arctic ice-cap gone, sea-levels risen by a metre or more, large swathes of the tropics uninhabitable or agriculturally useless, and corresponding geopolitical turmoil. 'Future generations', in other words, are already here, and one can look in their eyes and feel acute anxiety about what awaits them—and deep chagrin at one's own generation's collective failure to do anything serious about mitigating these prospects. And in the pang of recognising that appalling future awaiting some particular cherished innocent, there is now indeed something humanly intolerable.

So the question of what we can realistically hope for in the face of a rapidly and apparently inexorably deteriorating global climate situation, is now for me the question: what kind of hope can realistically be entertained for my new grandson's life?—bearing in mind that he was born in the year in which whatever hopes people had (maybe) placed in the Paris Agreement began clearly to unravel—with sabotage by the US looming, with its already becoming apparent that success merely on the terms of that agreement will not be enough to prevent warming going beyond +2°C, and with authoritative analyses showing that the chances of even so much success are receding rapidly (global emissions actually went *up* in 2017 for the first time since

2013[1]); and, on top of all that, with destabilising populist or nationalist forces adverse to Paris-style global co-operation ramping up worldwide.

And yet new life has to mean renewed hope. We could not go on living, in all the ways to which the constant busy production of new life is central, without such hope. But also, I stress again, we must be realists. It is not just that hopes unrealistically entertained will in no long time be defeated, or that seeing them defeated leads to recurrent activist burn-out and despair. Unrealistic hope actually sabotages possibility, wasting the very short time which we have left in which to act.

So where *are* we to find and lodge realistic hope? In this chapter I shall argue that quite ambitious hopes in the face of climate crisis still remain realistic—but also that we can easily misunderstand why they are so. Practically, you might say, that wouldn't matter, so long as we could go on working hopefully. But as a philosopher, I am committed to insisting that only a proper understanding will enable us to internalise such hopes *along with their limits*—that is, in their full, active strength.

Empirical realism and disillusion

Where to find hope? To pursue an answer to this question we need first to step back and ask a prior one about realism. What are we going to *count* as realism here? Because, as I started by suggesting, there are two fairly stark alternatives.

Let's start empirically. Human beings have a strong tendency—especially in broadly political contexts—to see and believe what they want to see and believe, or what the group or the Party want them to. This habit of acting on the basis of how we'd *like* things to be often results in painful subsequent collision with how things actually are; legendary and historical examples abound, from the sack of Troy through Napoleon and Russia to Brexit. So we have naturally evolved some fairly straightforward rules of thumb (if only we could stick to them) for aligning our picture of how things are with how they really are, before acting on it. These rules of thumb familiarly include paying less attention to what people *say* than to what they actually do, and particularly to what they have been in the habit of doing; assuming that, generally speaking, more immediate and self-interested concerns will win out over remoter and more altruistic ones; and generally, recognising the very wide purview of Sod's Law, that if something can go wrong, it usually will. This

---

1 'First CO$_2$ rise in four years puts pressure on Paris targets' http://www.bbc.co.uk/news/science-environment-41941265 (accessed 14.12.2017).

whole empirically-grounded interpretive mind-set, or apparatus of pragmatic suspicion, we might well call the *realism of disillusion*, since it draws on long experience of having been regularly disillusioned in order to puncture, as far as possible, further illusions going forward.

So what does such realism tell us about our present climate and ecological prospects? I think it now plainly tells us that if we *had* been going to prevent destructive climate change and its ecological consequences, we would have put in place genuine constraints on our carbon-emissions-generating behaviour worldwide quite soon after this first became a live issue, instead of dragging our collective feet from the 1980s through Rio 1992 to Paris 2015; but we didn't; and so, we're not going to prevent it. This little argument, which in *After Sustainability* (Foster, 2015: 2) I christened the 'vicious syllogism', is valid and its minor premise, at any rate, is plainly true. To the disillusioned eye, we palpably *didn't* do what was needed when we had the chance. We learnt to talk the talk over this period, but only ever to walk as much of the walk as would enable us to go on talking—as the painfully comic sequence of last-chance saloons along the way (from Copenhagen via Cancun to Paris…) makes plain enough. Meanwhile, the hypothetical major premise asserting that we are out of time looks as well-grounded in scientific evidence and hard-headed observation as any empirically-based counterfactual well could be.

The most effective way to demonstrate this last point is via the global carbon budget, the amounts of extra CO_2 emitted on top of what we have already put up there, to which we must limit ourselves if we want to retain a calculable chance of keeping atmospheric temperature rise within designated limits. The current scientific consensus is that at the time of writing (December 2017) we must live within a budget of between 770 and 800 billion more tons of emitted CO_2 in order to have a better-than-evens chance of keeping warming below 2°C—the level which has been taken by the climate policy community as just about manageable for global civilisation in something like its present form.

Eight hundred billion tons of CO_2 sounds quite a lot—enough, at any rate, to give us time for adaptation, for developing the cleaner technologies, carbon capture and storage systems and widespread substitution of renewables for fossil fuels which we will need either to keep within our carbon budget for the foreseeable future, or to move us into a zero-carbon regime when the budget runs out. Visit the Carbon Countdown website[2] maintained by *The Guardian*, however, and that impression of global elbow-room evaporates.

---

2   See https://www.theguardian.com/environment/datablog/2017/jan/19/carbon-count down-clock-how-much-of-the-worlds-carbon-budget-have-we-spent.

This webpage displays figures, changing in real time, for the total remaining carbon budget for a 66% chance of 2°C, and also for the period over which at current emission rates we will use this budget up. This latter figure is the most readily grasped, because only the seconds and minutes are changing as you watch: at the time of writing it was just over nineteen years. That is perhaps startlingly less than one might have expected, but the *really* shocking figure is the one for ongoing emissions. Confronting you at each visit to the site is a countdown clock showing the worldwide total in tons of CO_2 equivalent which has been emitted since you clicked in. As the numbers on this clock flicker past, so swiftly that you can only keep track of the mounting thousands, the scale of the challenge facing humanity emerges literally before your eyes. During several visits when writing this chapter, the counter was registering atmospheric carbon being added worldwide at the rate of 80,000 tons per minute, or more than 1,300 tons—a weight of carbon equal to that of around 105 new Routemaster buses—being pumped up into the skies *every second*. (That's some *seventy-five thousand* buses-worth since you started reading the chapter.) And when you remind yourself that atmospheric carbon dioxide is a *gas*, and therefore how much of it would be required to tot up to just a single ton in weight, and how far (for instance) you personally would have to drive in order to emit even that much, you can't help being overwhelmed by a sense of the gigantic scale, utter mad relentlessness and frenzied intensity at which human activity is spewing out this climate-destabilising pollutant around the globe.

And of course by the same token, these eye-glazing magnitudes indicate the scale of the changes which would be required to render all this oil-driven worldwide activity carbon-neutral by 2036, or 2045, or mid-century. Try to envisage that order of change, in all its geopolitical, economic, organisational, cultural and lifestyle dimensions, and it immediately becomes clear that revolution so all-encompassing and at such breakneck speed is without the faintest shadow of a precedent in human affairs. Nothing in any human experience suggests that change of this order could be practically possible.

Ah, but—someone might protest—what about the Paris Agreement? Hasn't that been the game-changer which gives us a better chance? But, as Brian Heatley and Rupert Read remind us in a Green House paper (Heatley and Read, 2017), achievement of the Paris goals depends entirely on the self-monitored meeting by each signatory country of its self-set emissions reduction targets. The apparent plausibility of these targets is largely an artefact of setting a modest-seeming annual reduction commitment against a target date which still feels comfortably distant—and we need only recall the racing carbon-counter to have the gravest doubts about *that*. But we shouldn't in any case need that reminder to make us realistically cynical about such a process, when entered upon by countries still uniformly committed to economic growth and rising material living standards, and therefore resembling nothing so much as a bunch of alcoholics left to agree and police their own safe drinking levels. Moreover, as climate scientists like Kevin Anderson (2012) had pointed out well beforehand, even if such a process inspired any confidence, the whole regime based on the 2°C target is still called into question by highly-likely feedback effects of which it takes no account, such as release of methane from thawing permafrost or increased heat absorption by the ocean as the Arctic ice melts. So even if the rogue Trump administration doesn't manage to wreck the whole Paris process, the clear implication of all this to a disillusioned eye is that global temperature rise will be at the very least 3–4°C over pre-industrial levels by 2100, and more likely 4–5°C—taking us into a world which Anderson describes as 'incompatible with any reasonable characterisation of an organised, equitable and civilised global community… [and] devastating for many if not the majority of ecosystems' (2012: 29)

So if that is realism, what can we realistically hope for, in terms of the life-conditions and life-chances of someone starting out in life, like my grandson, about now?

Just to recap: by around mid-century, massive feedback effects from our failure to control warming to 2°C will have started to become evident. These will arise specifically from methane release, the effects of rainforest dieback and ocean warming. Their consequences will entail sea-level rise of at least 3–4 feet with many low-lying coastal areas, including in Europe and the US, becoming uninhabitable; and simultaneously, dustbowl conditions in many areas currently heavily populated or intensively farmed, with resulting food and water insecurity leading to severe famine and drought in high-population third-world areas. While the broad parameters of these effects have been explored in scientific modelling, for example by the Royal Society (Anderson and Bows, 2011), they are individually unpredictable and could well be more severe than this suggests. (The general trend is for improved modelling in these fields to reveal previous estimates as underestimates). Even less predictable are the ways in which they will interact with each other, but plausibly to be anticipated on the basis of the historical record are proliferating wars over resources and territorial defence, at least some of them involving at least tactical nuclear weapons.

Is that prospect *hopeless*?—simply a cause for despair? Well, no—where life remains at all, no prospect is completely hopeless. But at best, it would seem, the hope can only be that anyone about whom one cares will be among the lucky minority. Those will be the people living in currently temperate zones which will become warmer but still remain habitable in the medium term. They will also need to be living outside urban areas (where life-support systems will collapse) and beyond the suburban wastelands (which will be stranded as transport infrastructure unravels), with some access to land which can be made productive, and within well-integrated, strongly led and practicably defensible communities.

Such a scenario is certainly something towards which it is possible to direct a *kind* of hope: it isn't yet a case of 'the living envying the dead', as after a nuclear holocaust, and it would be foolish to claim that life under these conditions—at the favoured end of the feasible—could never be worth having nor worth hoping for in default of anything better. But it would be hope only for the best of a very bad job, for a life pursued in temporarily tolerable local conditions against a grim background of cumulative global derangement. It would be hope for a stop-gap future on the way to nowhere, and that means a terribly diminished sense, not only of what a thriving human life requires, but of hope itself.

That is inescapable: the realism of settling for something of this order, *betrays* hope in the deeper sense in which it attaches to new life, to a newly arrived infant. The hope which is sprung in us by that arrival is not just that he or she, individually, will enjoy the best life practically possible; it is also a kind of impersonal hope, life-hope for what this newcomer embodies and represents, hope for the indefinite continuance of conditions in which human beings generally can lead flourishing human lives. Plainly, for the disillusioned realist, hope of *that* order for humanity as a whole looks as if it must now be abandoned.

The alternative: transformative realism

No wonder, then, that so many who can see this future looming and recognise it as dreadful strive to reassure themselves—to believe that something like the Paris accord can be made to deliver, or that a global 'sustainability' consensus can still be forged, or that heroic technology (which in practice would come down to untestable and wildly ambitious geoengineering—see the chapter by Paul and Read in this volume) will enable us to avoid or mitigate the more drastic consequences. But if what we have just been considering is indeed realism, such beliefs are themselves clearly little better than escapism. They can only be pursued as a form of denial.

Latterly, however, a different kind of response has been emerging—one which seeks to reclaim the idea of *realism* for an alternative stance towards what it nevertheless fully accepts. The keynote of this response sounds very clearly in these words of the American writer Rebecca Solnit, who has been a powerful advocate for it:

> Hope is not about what we expect. It is an embrace of the essential unknowability of the world, of the breaks with the present, the surprises.
> — Solnit (2004/2016: 109)

Disillusioned empirical observation picks up on what we have come to expect—the grindingly slow political foot-dragging, the failures of commitment, the widespread public resistance to any but superficial lifestyle shifts—and projects these forward. But as Joanna Macy and Chris Johnstone point out in the same spirit as Solnit:

> ...there is also *discontinuous* change...structures that appear as fixed and solid as the Berlin Wall can collapse or be dismantled in a very short time...a threshold is crossed...There is a jump to a new level, an opening to a new set of possibilities...
> — Solnit (2012: 189–91)

We might call this the realism of *transformation*, which of its nature is not predictable merely by extending forward the incremental history of any given situation. The concept of the tipping-point applies not only in climate science but here too: change which is quite implausible to all observers *ex ante*, the thought goes, can happen suddenly and startlingly once a critical mass of people starts to believe in it.

Writers like Solnit, Naomi Klein in *This Changes Everything* (2014) and George Monbiot in his *Out of the Wreckage: a New Politics for an Age of Crisis* (2017), document instances of this kind from the history of a variety of political and social movement campaigning. There is the case of the Berlin Wall already mentioned. Other examples, less sudden but no less dramatic in transformative extent, include the process which brought Nelson Mandela from nearly thirty years as a political prisoner of the apartheid regime to become the first black President of South Africa in only four years, or the seismic cultural shifts in perceptions of gender relations which have come about over the last few decades. Common to all such transformations, as both Solnit and Klein point out, was a more or less extended period during which small but dedicated groups of people took local-scale or even just symbolic action against apparently overwhelming odds—action which no plausible prediction at the time could have foreseen as likely to effect dramatic changes. And of course, this is just the sort of action which today is going on all over the world—against fracking, in pursuit of divestment from fossil fuel companies, in defence of indigenous people's rights against extraction or pollution, against unwarranted corporate power and influence wherever it manifests itself. We cannot know, it is claimed, any more than those whose cumulative small-scale commitments led to past changes could know, that all this diverse activity will not suddenly produce a transformative tipping-point.

Hence the inspirational rhetoric of Al Gore in his more recent film *An Inconvenient Sequel: Truth to Power:* 'This movement to solve the climate crisis is in the tradition of every great moral movement that has advanced the cause of humankind... We are very close to the tipping point beyond which this movement, like the abolition movement, like the women's suffrage movement, like the civil rights movement, like the anti-apartheid movement, like the movement for gay rights, is resolved into a choice between right and wrong'—the point at which the moral issues become so unignorably stark that entrenched habit ceases to be able to hold out against them.

It is on this kind of basis alone that the major premise of the vicious syllogism, which seems so compelling to the disillusioned eye, could be resisted, and nugatory gains over the past twenty-five years could cease to be decisive evidence for what may realistically be possible over the crucial next five or ten. Indeed, when we allow ourselves to think about what might emerge from diverse, networked social action in a set of dynamic, volatile, ethically-charged and systemically-linked contexts, what starts to look *un*realistic is the idea that we *could* reliably read off projections about the effects of our interventions from the course of events up to now

If commitment to the human capacity to break transformatively from the past is indeed realistic, then it opens the way for genuine hope—the full-hearted hope for a world in which human beings (and necessarily, therefore, a lot of other beings too) can go on flourishing. Indeed, such hope has itself to be the driving force of the necessary commitment, as what Macy and Johnstone (2012) call 'active hope': not just an attitude of desire directed at a possible future, but a conscious, structured practice of embarking oneself in working towards its realisation, against whatever apparent odds, sustaining that commitment by collaborative reinforcement and various kinds of spiritual discipline. As Solnit expresses the same idea: 'Hope locates itself in the premises that we don't know what will happen, and that in the spaciousness of uncertainty is room to act' (2004/2016: xiv).

Nor, to hope like this, do you have to be a professional climate activist, an ex-Vice-President on a mission or a campaigning journalist. Consider this passage from the distinguished climate scientist Kevin Anderson:

The current political and economic framework…seems to make [avoiding a breach of the critical 2°C threshold] impossible. But, it is not *absolutely* impossible. If [we] are prepared to make the necessary changes in behavioural and consumption patterns, coupled with the technical adjustments we can now make and the implementation of new technologies (such as low carbon energy supplies) there is still an outside possibility of keeping to 2°C.

— Anderson (2012: 38)

That's a pretty big *If*, when you consider what those necessary changes would have to involve. Yes, as Gore and many others regularly remind us, the daily influx of solar energy is enough to power the entire world economy many times over, and the technologies to capture and deploy this energy are rapidly becoming cheaper and more accessible. But to implement that technically-possible shift in time to avoid catastrophe, we would be looking, at minimum, for a decisive move away from neoliberal capitalist incentives, a very significant localisation of power to communities and away from global elites, and widespread transformation of consumerist lifestyles and expectations, which in turn would require significant real reductions in the energy-profligate transit of people and goods, with economic restructuring to match. Now Anderson, remember, though a powerful polemicist in this field, is first and foremost a research scientist, so his whole intellectual formation will emphasise the basing of projections and expectations firmly on carefully observed empirical evidence. But any application of that approach and method to the economic, organisational and sociological grounds for predicting change of that order of magnitude within anything like the coming decade, which is the relevant timescale, must surely have led him straight to disillusioned realism. And yet there he is, still advancing the 'outside possibility of keeping to 2°C'. So his '*If…*', and the hope which it sustains, must be grounded however tacitly in the alternative realism of transformation.

With this prompting, I think we may fairly characterise hope so entertained, and in general the kind of active hope that is premised on transformative potential, as explicitly *counter-empirical* hope. Anderson doesn't ask, as one would have expected a scientist of his calibre—or any scientist—to ask, just what probability the 'necessary changes' bear when weighed against the unprecedented nature of our current plight as it appears from all the brutal numbers. Rather he insists, if not quite in so many words and perhaps without consciously intending to (if his intention had become conscious, his scientific training in empirical observation would surely have rejected it), that *the past doesn't always show us the rules of possibility for the future*. Whatever may have happened up to now, and however it might pile up the odds against what we are hoping for, we can't know that there *won't* be a transformative leap to a new set of circumstances. Summoning our courage, therefore, we are to act in ways that keep that possibility open.

So now we can reframe the challenge—the vital question now for anyone who wants to go on committing themselves to serious climate activism: *is* counter-empirical hope, so based and encouraged, justifiable as realistic? Because if it isn't, of course, it can very rapidly become an indulgence—your hope can come to be more about *you* in the here and now, about what you can bear to accept or contemplate, than it is about those for whose future you are affecting to be concerned. And such indulgence will not only be an exercise in self-deception, it will also risk betraying us into wasting the very limited time now available on pursuing chimerical possibilities when we might be finding out what actual, solid reinforcements, albeit for a much harsher future, we could still hand on.

So: *is* it realistic?

Lessons from the past?

One way *not* to justify counter-empirical hope as realistic is to appeal to the evidence of history in the way its advocates characteristically do, as if past transformations somehow reduced the odds against us. That's at bottom a logical point, but as such it tends to get overlooked, so let me briefly flesh it out.

Consider a very recent such advocate, George Monbiot in *Out of the Wreckage*. His general line of argument in this new book aspires to revolutionary simplicity. Political prospects are transformed, he claims, only when people's minds are seized by a new narrative which makes sudden sense of their experience in a way that older, failing narratives have ceased to do. The new story which he offers is one of a huge silent majority of human beings who are by nature 'socially-minded, empathetic and altruistic' (neuroscience and environmental biology are supposed to *prove* this), but whose strong latent desire for a just and environmentally-responsible world is currently thwarted by a small minority global elite using 'lies and distraction and confusion' as their weapons. Mobilise the silent majority with a compelling project of restoration via a renewal of community and local belonging and the repossession of our common wealth, Monbiot contends, and 'there is nothing this small minority can do to stop us' (2017: 181).

This is inspiring Shelleyan stuff—'Ye are many, they are few!'—but how does he back it up? In the first place, by appealing to the way transformational change has characteristically happened hitherto: 'The great emancipations—from women's suffrage to Civil Rights, to independence from empire and the end of apartheid—came about through the mass mobilisation of citizens' (2017: 168). More specifically, he lays great emphasis on the recent example of what he calls Big Organisation, demonstrated in US Senator Bernie Sanders' 2016 Democratic Presidential nomination campaign. This was a networking process, in effect invented on the hoof, in which volunteer organisation was relied on to generate, exponentially, batteries of further volunteers each committed to speaking directly to a certain number of voters on the candidate's behalf. Inspired by Sanders' radical platform, a hundred thousand of these volunteers eventually managed to make contact with seventy-five million people—discovering in the process that actual conversation between activists and their fellow citizens can change deeply-held attitudes on controversial issues in ways which television advertisements or phone messages from a robot caller are much less effective at doing. (Well, who knew?) This example, he insists, helps us to start imagining how 'campaigns of any kind—not just to win elections but to win the battle over climate change...—can be transformed' (2017: 173–4).

Monbiot writes with eloquent passion, and when the climate stakes are as high as they are, it would be easy enough to get carried away. But it is only intellectual responsibility to note that he makes no move to *interrogate* these analogies anywhere in the book—whereas, if he wants them to support his line of argument, he is committed to interrogating them closely. This is because the thought informing any appeal to history to justify the reality of transformative possibility in the present must run like this: sometimes in the past, what seemed utterly implausible on the basis of experience up to that point has nevertheless actually happened—therefore, that X now seems to us utterly implausible on the basis of our own experience hitherto doesn't mean that it can't or won't happen. Or, as Solnit puts this: 'Studying the record... leads us...to expect to be astonished' (2004/2016: 109). A thought with this structure, however, needs to be handled carefully. You can argue that sometimes (in the past) what was then the past has been no reliable guide to what was then the future, and so what is now the past may sometimes similarly fail to be a reliable guide to what is now the future; but you can only do so on the assumption that *generally* the past *is* a reliable guide to the future—otherwise, of course, past astonishments wouldn't license us to expect future astonishment. But that general expectation of predictive reliability commits us to identifying *criteria of exceptionality* if we want to make claims about *particular* occasions when it might fail. We then have to say something like: in past situations with broad characteristics *a, b, c...*, the then past provided no reliable guide to the then future; our present situation vis-à-vis climate change has those characteristics; so it is a case in which we could reasonably 'expect to be astonished'. This reasoning, however, depends on interrogating one's analogies in the way that Monbiot avoids (as do all the others—I'm not just targeting him). And he does so advisedly, since not avoiding it would require close attention to *how like the past* in these respects the present situation actually is. That would have meant asking whether any of his large-scale examples of successful transformative change really offered a convincing precedent for overcoming the *un*precedented challenges now facing us. *Utterly* unprecedented: we are failing to deal with climate change largely because humanity has never before confronted a danger remotely like this—on this global scale, of this complexity, so insidiously interwoven with our lives and habits, and calling for this order of collective and personal insight, political flexibility, technical intelligence and cooperative restraint.

And does the ending of apartheid, for instance—the remedying of a glaringly visible injustice by a state which had been under international economic as well as moral pressure to do so for several decades—really have that much to tell us about what will be involved in remedying a pattern of injustice towards future people which everyone worldwide has long been under powerful economic pressure to keep firmly in place? Again, could waves of volunteers, even in their hundreds of thousands, persuade people into mould-breaking lifestyle change as easily as they could capture them for the simple, slogan-friendly objectives—'Break up the big banks', and so on—dictated to the Sanders campaign by the nature of mass-democratic politics? Different judgements are of course possible in answer to these and similar questions. But they would all have to involve weighing the force of the relevant analogy in relation to our current challenges and identifying the corresponding probabilities, and in any such assessment the probabilities as they present themselves to the eye of disillusioned realism would be bound to exert an enormous—surely, in fact, an irresistible—pull.

If, however, the response to *that* recognition is a further appeal to the ever-present possibility of unpredictable transformation in the face of any apparent unlikelihood, then it is plain that the historical analogies are doing no genuine work towards justifying our reliance on this kind of possibility. Those who urge us to invest ourselves in counter-empirical hope can't have it both ways. They can either rely on analogies from past experience, with all their disillusioning potential when we genuinely interrogate them, or they can treat what these analogies suggest as always unpredictably liable to transformative exception. What they can't do is rely on the analogies to *show* that they are always thus liable to exception.

Of course this doesn't mean that past triumphs can't be inspirational. To take another example from Gore's new film, we can certainly find inspiration for climate activism in Gandhi's movement to liberate India and his reliance on *satyagraha*, the 'force of truth' in the face of huge odds. But if it is realistic to be thus inspired, that is not because Gandhi's success in his strategy of speaking truth to power somehow reduces the odds stacked against us in our very different situation. It is because we can be inspired by his faith in it to recognise that *satyagraha* names a real transformational force in the world, a source of energy as objectively real as the sun, to which we too must then trust against whatever odds, and however *radically* different the forms of institutional and cultural oppression which we ourselves face, and the kind of human liberation we are now trying to achieve.

Keeping hope honest

Let me try to summarise my argument so far. In the face of anthropogenic climate change, and at the stage which we have now reached, it would be realistic on the evidence of how we have handled this so far to prepare for epochal human defeat—*unless* it is more realistic to trust, against the weight of any empirical evidence, to the open-ended creativity and transformative resourcefulness of the human spirit. But the claim that such trust and the hope founded in it are indeed more realistic, or realistic at all, cannot be *justified* in empirical fashion by appeal to the evidence of past transformations.

If not from history, then, how could we demonstrate that counter-empirical hope is realistic? Ultimately, I think, only by showing *reality to be such* that resolute human action can *always* create possibility: rather as I was just suggesting with *satyagraha*, or perhaps taking a cue from Hannah Arendt's claim that 'with the creation of man, the principle of beginning came into the world ...' (1958/1998: 177)—*ultimately*, that is, by work in existential metaphysics. But then, as a philosopher I probably would say something like that—and even as a philosopher, I fully recognise that metaphysics is not going to cut any significant mustard here. That is not just because metaphysics is abstract and remote while our plight is urgent and practical, nor even because I haven't space in this chapter for the complexity of argument required, but simply because no *argument* could persuade us to trust in our transformative creativity if we don't already. So my approach in what follows will be to change tack somewhat, and ask: given the counter-empirical nature of the hope which climate activists across the spectrum now seem to be explicitly avowing and embracing, what might still strike us as conducing to *un*realism, when that kind of hope is adopted as a principle of action for dealing with the world as we find it?—and how we could guard against its doing so?

Here of course I am not interested in the reply that it strikes us as unrealistic just because the odds against us are decisive; that is simply to fall back on the disillusioned stance. I want to enquire what, once we have committed to hoping against the odds, might still worry us about the practical realism of a counter-empirical faith in our ability to transform the world. This worry, I suggest, is going to be about dangers lying in the other direction from disillusion—that is, those of utopianism.

Everyone remembers Oscar Wilde in this connection: 'A map of the world that does not include Utopia is not worth even glancing at, for it leaves out the country at which Humanity is always landing.' (1973: 34) As Krishan Kumar glosses this, 'Utopian conceptions are indispensable...without them politics is a soulless void, a mere instrumentality without purpose or vision' (1991: 95–6). Utopias can serve to keep us aspirationally up to the mark, to ensure that at any rate we never set our sights on *less* than the best. Our approach here reflects the etymological ambiguity of the word, as between Ancient Greek *eu-topia*, a good place, and *ou-topia*, no-place—that is, as between the best possible outcome, and something *even better*—and utopia is an indispensable political and practical conception just because we frequently can't tell *ex ante* which of these categories whatever we happen to be aiming at falls into.

But then the difficulty is that in the domain of counter-empirical hope, there is no way for this crucial distinction between merely ambitious idealism and utopianism reliably to emerge. The value of utopian aspiration in practice depends entirely on our being able to correct continuously for the difference between the genuinely achievable and what goes beyond it, as experience reveals ongoingly which is likely to be which. That ongoing self-correction, however, is just what motivating ourselves by counter-empirical hope rules out, because *that* means we can always, at any difficult juncture, reassure ourselves that all we have encountered so far is failure to have reached the tipping-point of a transformative solution *yet*. This of course removes the check constituted by finding out what may be genuinely possible, albeit idealistic, on the basis of what actually happens. Aspirations which are in fact utopian are thus protected from challenge—since no apparent lengthening of the odds against them can *ever* rule out transformative change, which happens by definition against the odds.

The danger here is twofold. On the one hand, if utopia as incentive to idealistic aspiration ceases to be distinguishable from utopia as more than the best achievable, then when the latter is pursued unchecked and relentlessly enough, real possibilities, real opportunities and ultimately real lives, will be sacrificed to it. But there is also the subtler danger of vitiating the very quality of hope itself. Hope is, by definition, desire for an outcome believed to be neither certain nor impossible but somewhere in between. But for our hope to have any organic vitality, we must genuinely believe that such an outcome could *fail* to happen, and our commitment must therefore involve a risk and call for courage. And if we are protected, as above, against accumulating empirical evidence that what we are hoping for is actually utopian in the sense of lying beyond what is achievable, the line between this kind of commitment and playing false with ourselves becomes very thin indeed: and sooner or later we will step across it, from hoping for the outcome because we dare to insist on its still being possible, to refusing to concede its unlikelihood because our commitments, or our moral comfort, or our mental equilibrium, depend on our going on hoping. And hope so gerrymandered, factitious and essentially dishonest as it must be, will prove too brittle to support us as the demands made on it grow more and more exacting.

So the practical issue about realism becomes: how do we keep our indispensably counter-empirical hope *honest*? How do we keep it both invested courageously in transformative possibility, and at the same time in ongoing responsive touch with the reality where any transformation has to occur. The answer to which I want to point is: by recognising the human condition as inherently tragic. Tragic, that is, not just in the sense that death and mayhem are always happening, but in the sense which great works of drama and fiction have always shown to characterise our condition—where events are always liable to bring out destructive conflict inherent in countervailing commitments of evaluative agents.

What that means in the bluntest terms (and bracketing, again, the metaphysics of value which 'inherent' implies) is that in any real crunch, *there are no win-win situations*. That is why such recognition is a prophylactic against utopianism, which is always after the win-win outcome. But in the nature of our condition as evaluative agents, and in any affair that matters, there are only ever win-lose situations, where gain in terms of one value (real material benefit to presently-living people, say, or preserving the cultural integrity of settled local communities) involves loss, more or less grievous, in terms of one of the counter-values inevitably also in play (respecting the claims of future people, perhaps, or of refugees fleeing intolerable conditions). And such constantly conflicting values are characteristically incommensurable, so that there is also no 'winning on balance', either way—just pain and gain to be enjoyed and endured together, and lived beyond (for the survivors) in unguessably transformative ways.

I'll glance very briefly in conclusion at what applying that criterion—*no win-wins*—to the framing of our counter-empirical hopes might look like in practice.

So when Naomi Klein, for instance, writes that any plausible attempt to reduce global emissions properly 'will mean forcing some of the most profitable companies on the planet to forfeit trillions of dollars of future earnings by leaving the vast majority of fossil fuel reserves in the ground', and adds: 'Let's take for granted that we want to do [this] democratically and without a bloodbath' (2014: 452)—that's pretty clearly going for a win-win. We almost certainly won't be so lucky, nor, if we are taking our tragic exposure seriously, should we build the hubristic demand that we be so lucky into our plans of action as any sort of bottom line.

Someone might object to this that we have then merely gone around in a big circle, because accepting the inevitability of bloodbaths and less-than-democratic governance is no different from admitting defeat as disillusioned realists and brutally scaling back our hopes. But I think the crucial difference is actually still very much in play: the difference between giving up because the odds against us show that we have run out of possibility, and fighting on in active hope which *creates* possibility. It can't, however, create limitless possibility—and if the limits aren't going to be encountered and negotiated empirically, because of the structure of counter-empirical hope, we must respect them through due acknowledgement of how our agency is *inherently* tragically configured.

Again, such acknowledgement probably involves accepting that 'sustainable prosperity', a policy pitch much bandied about lately,[1] is going for a win-win—at least if *prosperity* means anything in touch with what the large majority of Western people are going to understand by it. (And if it doesn't, if it is just a marker for 'whatever lifestyle actually turns out to meet the constraints of sustainability', then the phrase amounts to no more than a rather blatant attempt to pull the wool.) We can, maybe, recover a world in which a wide diversity of human lives across the globe can be both conducted sustainably and *worth living*, but only at the end of a very painful process of adjustment and downsizing—and to plan for a world in which everyone (or even *anyone*, much) 'prospers' in the securely-established abundance which that word now inescapably suggests to Western ears, would be only another exercise in hubris.

We can realistically invest our energies in work sustained by counter-empirical hope, I am suggesting, provided that we do so within the framework

---

1 See for instance the work of the Centre for the Understanding of Sustainable Prosperity, https://www.cusp.ac.uk/.

of a *tragic imaginary*. An imaginary (see Earle, 2017 for the concept) is a horizon of possibilities framing all our aspirations. Here, the framing has these
aspirations as, always, tragically vulnerable. We can go on actively hoping
and working for radical changes in our society, economy and settled habits
which a disillusioned realism would by now have given up on, but only in that
resolutely anti-utopian spirit. We can pursue, say, a basic income scheme—but
not the elimination of poverty, here or anywhere; effective carbon taxes, but
not a worldwide regime of contraction and convergence; significant economic
localisation, but not equality of either opportunity or outcome across all local-
ities (differentiation really does mean *difference!*); serious re-socialisation of
key infrastructural capital, but not, alas, the overthrow of capitalism; humane
international arrangements for addressing climate-driven migration, but not,
appallingly, the avoidance of drought and famine across the planet...and
so on.

More generally, we can maybe still hope to avoid climate catastrophe—but
no longer, surely, to avoid what will impact on us over the coming decades as a
series of increasingly grievous disasters. It is in the space between disaster and
catastrophe that honest hope must now be nourished.

CODA—WHERE NEXT?

John Foster on behalf of the Green House collective

Any book offering to explore and analyse complex socio-political issues must expect to raise more questions than it answers. The present book is no exception. But it belongs to our subject matter that this unfinished business must be understood in a very particular way.

Facing up to climate reality is of its nature a *starting point*—something which we have only really achieved, if that then leads on in unguessable directions. For it entails recognition that adapting to the new conditions which are coming must be *radically* transformative. That is, adaptation must not only reach beyond incremental change, within the framework of existing systems and structures, to challenge and recast that framework itself—the sense in which the IPCC uses the term 'transformational adaptation' (IPCC, 2014a: 189)—but even more demandingly, it must involve letting go of the idea that we can maintain control of that process of recasting, that we can plan its direction and manage its results. Along with residual aspirations to continue onwards down the familiar track of qualified material progress or 'sustainable prosperity', we will have to abandon the whole framing of our condition as *problematic*, with (somehow) 'solutions' to be found. The reality of dangerous anthropogenic climate and ecological disruption which we must now confront involves far too much dynamic instability and unpredictability for any further indulgence of such Enlightenment self-flattery. Acknowledging our real situation now means embarking ourselves in forms of ongoing resilient coping (economic, social, psychological, ethical and existential) which will need to be *open-endedly* flexible and adaptable.

It follows that the relation between what the foregoing chapters attempt and what would be involved in taking this work forward cannot just be a matter of further spelling out things already implied or hinted. What we have left unsaid about the ways in which transformative adaptation might be realised in practice is not what we have simply lacked room or time for saying, but what *cannot yet* be said because it remains to be revealed—or rather, created.

193

Seen in that light, each of these chapters has been seeking to indicate what 'facing up' might look like in each respective topic area—what the points of departure for transformative adaptation would need to be, for it to be claimed that we had genuinely confronted our real situation. We might say, with an eye on the Brexit process (which for all its ugliness surely has things to teach us about moving into uncharted political and institutional territory), that they lay down clear red lines, to which such adaptation must require us to conform, whatever drastic social, economic and attitudinal shake-outs might result. But of course, the Brexit 'red lines'—no more free movement, no customs union and so on—represented an ideological interpretation of what the ill-conceived 2016 referendum was supposed to have meant, while ours are unambiguously demanded by respect for the planetary boundaries (Rockstrom et al., 2000) against which humanity is now so clearly pushing. It would be better to emphasise their contrasting creativity and the politics whence they originate by calling them instead *green* lines—ways of demarcating basic minima for the kind of (necessarily demanding and disruptive) collective action in which hope could still be invested, and from which painful life-renewal might yet spring.

What, then, are the green lines, in this sense, which stand out retrospectively from the book?

Part I, it will be recalled, deals in those which must configure any hopeful politics. For Richard McNeill Douglas, abolishing the systemic drive towards accumulation characteristic hitherto of the dominant form of capitalism represents the central requirement for an economy which both respects environmental limits and retains the plurality and freedom of markets and private initiative. Peter Newell argues that some form of contraction-and-convergence model looks like the inescapable framework for a global regime which apportions remaining 'ecological space' either justly or (in the not very long run) viably. Rupert Read and Kristen Steele emphasise that hopes for human flourishing to be retrieved out of climate disaster turn on what ordinary people may be able to bring out of extraordinary situations, and call for a concerted shift of our political and economic arrangements towards facilitating localisation both in anticipation and in response—which would be transformative adaptation already in action.

In Part II, where we consider fundamentals for systems capable of bearing up under increasing climate stress, Jonathan Essex argues that stopping the juggernaut of further carbon-intensive urbanisation means ensuring that land-use planning, emissions reduction and support for community resilience become an explicitly integrated process; this implies among other things making it a clear green line for any continuing capitalism that it should decarbonise society. Anne Chapman's quietly scary case studies of dangerous weather events flag up above all the need to *defragilise* our enabling arrangements for everyday life, for example electricity supplies and structures of support for elderly people, so that they have enough inherent robustness to cope with the oncoming normalisation of the extreme. At a more general level, Helena Paul and Rupert Read demonstrate that the precautionary principle must become the recognised basis of all policy and planning, not just of our approach to (supposedly) managing the stratosphere, if we are to resist temptations towards overweening techno-hubris which will only intensify as climate disasters loom larger.

Green lines are equally applied, in Part III, to the intellectual, psychological and existential framing of these issues. Brian Heatley seeks to provide against the different form of hubris which consists in dismissing the past, however distant and different, as offering nothing from which we could learn. Nadine Andrews and Paul Hoggett underline how those who face the facts of climate and ecological crisis may experience feelings of loss, guilt, anxiety, shame or despair that can be very hard to bear. How these feelings are dealt with shapes how people respond to the crisis, so we shall need psychological resources that can help build resilience and support positive change at individual, community and societal levels. My own chapter points out, however, than such resources can only ever offer partial help. The very bottom line (still, paradoxically enough, a green one) is that the human condition itself entails irresolvable value conflict, so that whatever is won goes always with grievous loss, again and again, indefinitely—and recalling this long-standing tragic truth is a requirement for staying in touch with reality not only in the climate context, but everywhere.

If those are indeed our key green lines, what does that mean going forward?

It means, we take it, that those impressed by our argument should be engaging in a programme of collaborative exploration with people in every relevant sector, to explore in detail what respecting those fundamentals would now mean, what manifestations of radical shake-out it might be likely to produce, and how far those can realistically be anticipated and prepared for. And by the relevant sectors, we don't just intend the obvious candidates—groups of environmental activists, Green politicians and the like. While these might be expected to be sympathetic, the project of facing up to climate reality also needs to be reaching out more ambitiously to teachers, scientists, civil servants, lawyers, manufacturers, service industry managers, security professionals and anyone else whose role in reinventing the possibilities of civilisation will become crucial as climate reality closes in on us.

Such a programme must indeed cast its net even more widely than this suggests. Of particular importance will be including in exploratory dialogue those who can speak authoritatively for the economically poor or otherwise vulnerable, whether in Britain and similar societies or in the global South. For these constituencies, some versions of radical shake-out could mean the difference not between historically anomalous levels of material affluence and rougher forms of coping, but between life and death—and it is implicit in all our green lines that such outcomes be minimised where they can be. Despite work which Green House itself has done on the ideas of Just Transition and 'climate jobs',[2] that will certainly be a conversation stripped bare of comforting reassurance, but it must be engaged in.

Equally important to factor into such dialogue is the danger that focussing almost exclusively on the existential threat of climate change, as we have done in this book, might lead us to overlook the terrifying biodiversity crisis—the sixth great extinction which humanity is so far driving on even more by direct habitat-destruction than through changing the climate. We must ensure that transformative adaptation, however it develops, fully recognises the imperative for its green lines to apply here too.

Promoting such a programme as just sketched means, for us in Green House, using this book and any accumulating spin-offs in a carefully organised way as tools to spur and facilitate, but by no means to constrain or prejudge, the necessary exploration. Nor is that, of course, just a project for Green House, but something for which any reader whom we have persuaded can—and we hope will—use the book as an incentive in his or her own sphere of action.

It might be queried why, given our analysis of the plight in which humanity now stands, we still think such dialogical work—such commitment to reason and shared understanding as well as honesty—to be worthwhile. We are encouraged in this view by the extent to which, since we first conceived of the project three years ago, confrontation with climate reality seems to have come out of the closet. Aldo Leopold regretted, a full seventy years back (1949: 19), that the grim task of the ecologist was to see 'the marks of death in a community that believes itself well and does not want to be told otherwise'—and for most of the period since, that has remained broadly true. But some tectonic plates are now shifting, and the community's progressivist complacency is at last beginning to falter. This book is appearing at a moment which, amidst the dire straits in which our species has now placed itself, genuinely contains a modicum of propitiousness. The rise of Extinction Rebellion[3] is perhaps the most striking such straw in the wind. What is so refreshing about this

---

2 See https://www.greenhousethinktank.org/climate-jobs.html.
3 See https://rebellion.earth/.

movement—and what may partly explain its extraordinarily rapid growth and impact—is its refusal to abide to by the norms of mainstream 'climate communication', its raw and powerful message of grief, anger and quite-likely-doom. It is a movement which consciously forgoes the cheaper optimisms of most earlier campaigns for social change, an uprising perhaps more of courage and passion even than of hope—which is, paradoxically, what from our perspective is the most hopeful thing about it.

Nor is Extinction Rebellion quite alone. Though less dramatic, essentially similar kinds of demonstration are now starting to emerge. A very heartening one is the 'school strike' initiative, whose very articulate young Swedish instigator made such an impression at the Katowice climate summit.[4] This has spread to, Australia, Belgium, the US and Japan, and latterly to Britain, showing how young people across the world are increasingly clear-eyed and fiercely indignant about our real situation.

Of a different order, but maybe even more significant, there are the growing manifestations of imaginative confrontation with these issues. In serious writers of the calibre of Cormac McCarthy, Ian McEwan, Barbara Kingsolver and Margaret Atwood, and in some of the better artwork curated by the Dark Mountain project,[5] awareness deepens that our starkly unprecedented climate and ecological jeopardy is rapidly becoming the fundamental life-issue of our times. (The absence of anything on these developments is perhaps the most regrettable omission from our book as it stands.) In popular culture too, where these matters have scarcely figured to date, forms such as the cinema are beginning to address them.[6] Certainly, a key green line among others must be explicit recognition that the creative imagination in all its varieties can tell us at least as much as any form of science about the human condition. Acknowledging that, we can also remember that it has always deeply shaped human possibility.

So there are signs that climate honesty is starting to break through. Are we, for all that, too sanguine? Some, like the recently-formed Rapid Transition Alliance,[7] claim to find 'evidence-based hope' for the possibility of the needed transformations. This can be persuasive when it means, for instance, pointing to what happened when the Icelandic volcano erupted in 2010 to demonstrate

4 See https://www.theguardian.com/environment/2018/dec/04/leaders-like-children-sc hool-strike-founder-greta-thunberg-tells-un-climate-summit.

5 See https://dark-mountain.net/. See also Cape Farewell https://capefarewell.com.

6 See on this Rupert Read's new book *A Film-Philosophy of Ecology and Enlightenment* (Read, 2018); details at https://www.routledge.com/A-Film-Philosophy-of-Ecology-and -Enlightenment/Read/p/book/9781138596023.

7 See https://www.rapidtransition.org/about/.

that we can, at a pinch, actually live without ubiquitous budget air travel. But when it suggests that we can also find reassurance in the success of past popular movements for change, one must be sceptical for the kind of reason explored in some detail in Chapter 9 above. The challenge facing effective action against dangerous climate change is surely many orders of magnitude greater than that which faced activists for civil rights, say, or for feminism. What we need now, as this whole book has amply shown, is not just to remedy particular, even systemic, abuses, but to achieve sweeping change, from the bottom up and right across politics, economics, culture, morality and psychology. That must call for hope of a different order—hope which is in Jonathan Lear's (2006) terms *radical*, which is still embraced when the empirical and cultural grounds for it seem absent.

This is the most complex and exacting challenge which humanity has ever faced. To live in these times and to face their reality means risking grief and taking responsibility in the pursuit of uncertainty. Are the radical transformations that might avert catastrophe really possible? Only time will tell. Meanwhile, in the spirit of this book, we go on: neither in optimism nor in easy hope, but in the courage required for that more difficult kind of hope, if it is to be found.

BIBLIOGRAPHY

2017 Atlantic Hurricane Season. 2018. https://en.wikipedia.org/wiki/2017_Atlantic_hurricane_season. Accessed 11 May 2018.

2017 South Asian Floods. 2018. https://en.wikipedia.org/wiki/2017_South_Asian_floods. Accessed 11 May 2018.

Addis, B. 2008. "Briefing: Design for deconstruction." in *Proceedings of the Institution of Civil Engineers: Waste and Resource Management*. Vol.161:1, 9–12. London: Thomas Telford.

Allen, Myles et al. 2018. *Global warming of 1.5°C. An IPCC Special Report on the impacts of global warming of 1.5°C above pre-industrial levels and related global greenhouse gas emission pathways, in the context of strengthening the global response to the threat of climate change, sustainable development, and efforts to eradicate poverty*. IPCC.

American Psychological Association. 2009. *Psychology & Global Climate Change: addressing a multifaceted phenomenon and set of challenges*. Washington DC: American Psychological Association.

Anderson, K. 2012. "Climate change going beyond dangerous – Brutal numbers and tenuous hope" *Development Dialogue* September 2012 http://www.grandkidzfuture.com/occasional-pieces/ewExternalFiles/Anderson%20Going%20beyond%202012.pdf (accessed 9 August 2017)

Anderson, K. 2013. Blog at http://kevinanderson.info/blog/avoiding-dangerous-climate-change-demands-de-growth-strategies-from-wealthier-nations/ (accessed 24 November 2016).

Anderson, K. 201.5 "Duality in climate science", *Nature Geoscience* 8, 898–900.

Anderson, K. 2016. "The hidden agenda: how veiled techno-utopias shore up the Paris Agreement". *Kevin Anderson blog*. 6th January. Available at: https://kevinanderson.in fo/ blog/the-hidden-agenda-how-veiled-techno-utopias-shore-up-the-paris-agreement/ (Accessed 21 December 2018).

Anderson, K. and A. Bows 2011. "Beyond 'dangerous' climate change: emission scenarios for a new world" *Philosophical Transactions of the Royal Society A*, 369, 20–44.

Anderson, K. and A. Bows-Larkin. 2012. "Executing a Scharnow turn: reconciling shipping emissions with international commitments on climate change." *Carbon Management*. Vol.3:6 (2012): 615–628. https://www.tandfonline.com/doi/full/10.4155/cmt.12.63.

Anderson, K. and G. Peters 2016. "The trouble with negative emissions" *Science* Vol. 354, Issue 6309, pp. 182–183.

Andrews, N. 2016. "Bill McKibben is wrong: humans and nature are not 'at war'" cultureprobe, August 18. https://cultureprobe.wordpress.com/2016/08/18/bill-mckibben-is-wrong-humans-and-nature-are-not-at-war/

Andrews, N. 2017. "Psychosocial factors influencing the experience of sustainability professionals." *Sustainability Accounting, Management and Policy Journal* 8: 445–469.

Arendt, H. 1958 /1998. *The Human Condition* (2nd ed.) Chicago: University of Chicago Press.

Atkins, UCL and DFID. 2015. *Future Proofing Cities: Risks and Opportunities for Inclusive Urban Growth in Developing Countries.* Epsom: Atkins. http://www.atkinsglobal.co.uk/~/media/Files/A/Atkins-Corporate/group/services-documents/future-proofing-cities/fpic-indian-cities-report-final-report-web.pdf.

Bailey, I., A. Gouldson and P. Newell 2011. "Ecological modernisation and the governance of carbon: A critical analysis", *Antipode* 43(3): 682–703.

Baoxing, Qiu. 2007. *Harmony and Innovation: Problems, Dangers and Solutions in Dealing with Rapid Urbanisation in China.* Italy: L'Arca Edizioni.

Barrett, Adam B. 2018. "Stability of Zero-Growth Economics Analysed with a Minskyan Model". *Ecological Economics* 146 (April): 228–239. doi:10.1016/j.ecolecon.2017.10.014.

Bartlett, M. 2017. *Community Emergency Planning in Lancaster.* Presentation to 'Dealing with Extreme Weather' conference, Lancaster, 28 October. Available at https://www.greenhousethinktank.org/dealing-with-extreme-weather.html.

Bateson, G. 1982. *Steps to an Ecology of Mind.* Reprint 1987. Northvale, New Jersey: Jason Aronson.

Baumol, William J., Robert E. Litan, and Carl J. Schramm. 2007. *Good Capitalism, Bad Capitalism, and the Economics of Growth and Prosperity.* London: Yale University Press.

BBC. 2007. *Disasters – extreme heat kills thousands in Paris* available at https://www.youtube.com/watch?v=Hx2940wcooo&feature=youtube.

BBC. 2018. *Political Electricity.* Analysis, Radio 4, broadcast on 19/2/18. Available at https://www.bbc.co.uk/programmes/b09rx4z9

Beerling, David et al. 2018. "Farming with crops and rocks to address global climate, food and soil security". *Nature Plants.* 4: 138–147.

Bellofiore, Riccardo, ed. 2013. *Rosa Luxemburg and the Critique of Political Economy.* London: Routledge.

Benjamin, Walter 1940 [2005]. *On the concept of history.* Available at: https://www.marxists.org/reference/archive/benjamin/1940/history.htm (accessed 20 December 2018).

Best Foot Forward. 2002. City Limits London. Accessed December 19 2018. www. citylimitslondon.com/downloads/Execsummary.pdf.

Biermann, F. 2014. *Earth Systems Governance: World Politics in the Anthropocene* Cambridge MA: MIT Press.

Bio-Based News 2018. "CO2 use is slowly picking up speed". 28th March. Available at: http://news.bio-based.eu/co2-use-is-slowly-picking-up-speed/ (Accessed 21 December 2018).

Bion, W. 1962. *Learning From Experience*. London: Heinemann.

BioRegional. 2009. *Capital Consumption Report*. Sutton: BioRegional. https://www. bioregional.com/capital-consumption-report/.

Blackwater, Bill (R.M.Douglas). 2015. "Rediscovering Rosa Luxemberg". *Renewal: A Journal of Social Democrac* 23 (3): 71–85.

Blauwhof, Frederik Berend. 2012. "Overcoming Accumulation: Is a Capitalist Steady-State Economy Possible?" *Ecological Economics* 84 (December): 254–261. doi: 10.1016/j.ecolecon.2012.03.012.

Blewitt, J. and R. Cunningham (eds.) (2014) *The Post-Growth Project.* London Publishing Partnership.

Boonstra, Wiebren J., and Sofie Joose. 2013. 'The Social Dynamics of Degrowth'. *Environmental Values* 22 (2): 171–189. doi:info:doi/10.3197/096327113X135815 61725158.

Brown, O., Hammill, A., & McLeman, R. 2007. "Climate change as the 'new' security threat: implications for Africa". *International Affairs, 83*(6), 1141–1154.

Brunner, H.P. 2013. *What is Economic Corridor Development and What Can it Achieve in Asia's Subregions?* Accessed December 19 2018. https://www.adb.org/publications/economic-corridor-development-and-what-it-can-achieve-in-asia-subregions.

Bull, H. 1977. *The Anarchical Society* London: Macmillan.

Bumpus, A. and D. Liverman 2008. "Accumulation by de-carbonization and the governance of carbon offsets". *Economic Geography* 84(2): 127–155.

Burke, A. et al. 2016. "Planet Politics: A Manifesto from the End of IR", *Millennium: Journal of International Studies* 44, no. 3 (2016): 499–523.

Burnham, P. 1990. *The Political Economy of Postwar Reconstruction*. London: Macmillan.

Buxton, N. and B. Hayes 2016. *The Secure and the Dispossessed: How the Military and Corporations are Shaping a Climate-Changed* World London: Pluto Press.

Cahen-Fourot, Louison, and Marc Lavoie. 2016. "Ecological Monetary Economics: A Post-Keynesian Critique". *Ecological Economics* 126 (June): 163–168. doi:10.1016/j. ecolecon.2016.03.007.

Carrington, Damian 2018. "UK to miss legal climate targets without urgent action, official advisers warn". *The Guardian.* 17th January. Available at: https://www.theguardian.com/environment/2018/jan/17/uk-to-miss-legal-climate-targets-without-urgent-action-official-advisers-warn (Accessed 21 December 2018)

Carrington, Damien. 2018. "Gulf Stream current at its weakest in 1,600 years, studies show." *The Guardian,* 28th April. https://www.theguardian.com/environment/2018/apr/11/critical-gulf-stream-current-weakest-for-1600-years-research-finds.

CB Insights. 2018. *Oil And Gas Corporates Are Investing In Clean Tech, Analytics, And The Internet Of Things.* Accessed December 19 2018. https://www.cbinsights.com/research/oil-gas-corporate-venture-capital-investment/.

Chandler, D. et al 2017. "Anthropocene, Capitalocene and Liberal Cosmopolitan IR: A Response to Burke et al.'s 'Planet Politics'". *Millennium: Journal of International Studies* 1–19.

Clayton, S., P. Devine-Wright, P. C. Stern, L. Whitmarsh, A. Carrico, L. Steg, J. K. Swim, and M. Bonnes. 2015. "Psychological Research and Global Climate Change." *Nature Climate Change* 5 (7): 640–46.

Climate Code Red 2017, Online at http://www.climatecodered.org (accessed 6 March 2017).

ClimateActionTracker 2015. Online at http://climateactiontracker.org/indcs.html (accessed 13 December 2015).

Cohen, B., 2006. "Urbanization in developing countries: Current trends, future projections, and key challenges for sustainability." *Technology in Society*, 28(1–2), pp.63–80. https://www.sciencedirect.com/science/article/pii/S0160791X05000588.

Connelly, James, and Graham Smith. 2002. *Politics and the Environment: From Theory to Practice.* 2nd ed. London: Routledge.

Corfee-Morlot, J., Marchal, V., Kauffmann, C., Kennedy, C., Stewart, F., Kaminker, C., and Ang, G. 2012. *Towards a Green Investment Policy Framework: The Case of Low-Carbon, Climate-Resilient Infrastructure.* OECD Environment Working Paper 48. Paris: OECD. http://dx.doi.org/10.1787/5k8zth7s6s6d-en

Coser, L. 1961. "Introduction". In *Max Scheler: Ressentiment.* New York: The Free Press.

Coumou, D., Di Capua, G., Vavrus, S., Wang, L., Wang, S. 2018. "The influence of Arctic amplification on mid-latitude summer circulation", *Nature Communications.* DOI: 10.1038/s41467-018-05256-8. Accessed 21 August 2018.

Cramer, P. 1998. "Coping and Defense Mechanisms: What's the Difference?" *Journal of Personality* 66 (6): 919–946. doi:10.1111/1467-6494.00037.

Crompton, T., and T. Kasser. 2009. *Meeting Environmental Challenges: The Role of Human Identity.* Godalming, UK: WWF-UK.

D'Alisa, G., Demaria, F. and Kallis, G., 2014. *Degrowth a Vocabulary for a new Era*, Routledge, London.

Daly, Herman E. 1991. *Steady-State Economics: Second Edition With New Essays.* Washington, DC: Island Press.

Daly, Herman E. 2010. "The Operative Word Here Is 'somehow'". *Real World Economics Review* 54 (27 September): 103.

Daly, Herman, and Benjamin Kunkel. 2018. 'Ecologies of Scale'. *New Left Review*, II,, no. 109: 81–104.

Dante 2003, *The Divine Comedy*, London, Penguin.

Dasgupta, S., Gosain, A. K., Rao, S., Roy, S., and Sarraf, M. 2013. "A megacity in a changing climate: the case of Kolkata". *Climatic Change*, Vol.116:3–4 (2013): 747–766. https://link.springer.com/article/10.1007/s10584-012-0516-3.

Dercon, S. 2011. *Is Green Growth Good for the Poor?* Policy Research Working Paper 6936. Washington: World Bank. https://openknowledge.worldbank.org/bitstream/handle/10986/18822/WPS6936.pdf?sequence=1.

Diamond, J. 2005. *Collapse: how societies choose to fail or succeed.* New York: Viking Press.

Dietz, S., and N, Stern. 2015 "Endogenous growth, convexity of damage and climate risk: how Nordhaus' framework supports deep cuts in carbon emissions." *The Economic Journal* Vol.125:583 (2015): 574–620.

Displacement Solutions. 2012. *Climate Displacement in Bangladesh: The Need for Urgent Housing, Land and Property Rights Solutions.* Geneva, Switzerland: Displacement Solutions. http://displacementsolutions.org/wp-content/uploads/DS-Climate-Displacement-in-Bangladesh-Report-LOW-RES-FOR-WEB.pdf.

Dobson, A. 2014. *The Politics of Post-Growth.* Dorset: Green House.

Doherty, T.J. and S. Clayton. 2011. "The psychological impacts of global climate change." *American Psychologist*, 66(4): 265–276.

Doig, A., and J, Ware. 2016. *Act Now: Pay Later. Protecting a billion people in climate-threatened coastal cities.* London: Christian Aid. https://www.christianaid.org.uk/sites/default/files/2017-08/act-now-or-pay-later-protecting-1-billion-people-climate-threatened-coastal-cities-may-2016.pdf

Doppelt B. 2016. *Transformational Resilience: How Building Human Resilience to Climate Disruption can Safeguard Society and Increase Well-being.* London: Routledge.

Dowd, D. 1989. *The Waste of Nations* London: Westview Press

Earle, S. 2017 "Imaginaries and Social Change", *Medium*, https://medium.com/@samraearle/imaginaries-and-social-change-2e0c8c093c25 (accessed 10 August 2017)

EAS AC (European Academies Science Advisory Council). 2018. *Extreme weather events in Europe: Preparing for climate change adaptation: an update on EASAC's 2013 stu dy.* Online at https://easac.eu/fileadmin/PDF_s/reports_statements/Extreme_Weather/EASAC_Statement_Extreme_Weather_Events_March_2018_FINAL.pdf

Eckersley, R. 2004. *The Green State* Cambridge, MA, MIT Press.

EEA 2001 *Late Lessons From Early Warnings: The Precautionary Principle 1896–2000*, Environmental Issues Report No 22, European Environmental Agency (Copenhagen).

Essex, J. 2014. *How to Make Do and Mend Our Economy: Rethinking Investment Strategies for Construction and Industry to meet the Challenge of Sustainability.* http://www.green-housethinktank.org/uploads/4/8/3/2/48324387/make_do_and_mend_inside2_small.pdf (Accessed 19 December 2018.)

Essex, J. and C. Gallego-Lopez. 2014. *Understanding the Relative Strength of Climate Signals Compared to Other Expected Development Results.* London: DFID. https://www.gov.uk/dfid-research-outputs/rapid-desk-based-study-understanding-the-relative-strength-of-climate-signals-compared-to-other-expected-development-results

EU 2015. *European Commission identifies the infrastructure priorities and investment needs for the Trans-European Transport Network until 2030.* https://ec.europa.eu/transport/themes/infrastructure/news/2015-01-15-corridors_en. (Accessed 19 December 2018.)

Evans, A.W. 1972. "The Pure Theory of City Size in an Industrial Economy." *Urban Studies*, 9(1), 49–77. UK: Sage. https://journals.sagepub.com/doi/abs/10.1080/00420987220080031.

Fairhead, J., Leach, M. and Scoones, I. 2012. "Green grabbing: A new appropriation of nature?" *Journal of Peasant Studies*, 39(2): 285–307.

Farand, C. 2018. "Extinction Rebellion eyes global climate campaign of non-violence" *Climate Home News* 20th November, Available at: http://www.climatechangenews.com/2018/11/20/extinction-rebellion-eyes-global-climate-campaign-non-violence/

Fonagy, P., G. Gergely, E.L. Jurist, and M. Target. 2002. *Affect Regulation, Mentalization and the Development of the Self.* New York: Other Press.

Foroohar, R. 2016. "Crisis in Capitalism and the Role of Wall Street." in *The Guardian* May 21st, 2016. https://www.theguardian.com/commentisfree/2016/may/21/crisis-in-capitalism-and-role-of-wall-street.

Fossier, Robert, (ed.) 1986. *The Cambridge Illustrated History of The Middle Ages, Vol 3,,* Cambridge, CUP.

Foster, J. 2015. *After Sustainability: Denial, Hope, Retrieval* Abingdon: Earthscan from Routledge.

Foster, J. 2017a (ed) *Post-Sustainability: Tragedy and Transformation* Abingdon: Routledge.

Foster, J. 2017b 'On letting go' in Foster 2017a.

Foster, J. 2017c. *Deep Hope: Climate Tragedy, Realism and Policy.* http://www.greenhouse thinktank.org/uploads/4/8/3/2/48324387/towards_deep_hope_inside_final.pdf.

Frantz, C.M, F.S. Mayer, C. Norton, and M. Rock. 2005. "There is no I in nature: the influence of self awareness on connectedness to nature." *Journal of Environmental Psychology.* 25(4): 427–436.

Freestone, R. 2009 "Planning, Sustainability and Airport-Led Development." *International Planning Studies.* 14.2 (2009): 161–176. https://rsa.tandfonline.com/doi/abs/10.1080/13563470903021217

Friedman, B. 2006. *The Moral Consequences of Economic Growth* London: Vintage Press.

Fritsche, I. and K. Häfner. 2012. "The malicious effects of existential threat on motivation to protect the natural environment and the role of environmental identity as a moderator." *Environment and Behaviour.* 44(4): 570–590.

Fritz, Charles 1996. "Disasters and mental health: Therapeutic principles drawn from disaster studies" in *Historical & Comparative Series.* 10. Disaster Research Centre.

Fulcher, James. 2004. *Capitalism: A Very Short Introduction.* Oxford: Oxford University Press.

Funtowicz, Silvio and Jerome Ravetz 1993. "Science for the post-normal age". *Futures.* 31(7): 735–755.

(Accessed December 19 2018).

Galofré-Vilà 2017, Gregori Galofré-Vilà, Andrew Hinde and Aravinda Guntupalli, *Heights across the last 2000 years in England,* University of Oxford, Discussion Papers in Economic and Social History, Number 151, January 2017.

Gass, Henry 2013. "Geoengineering could weaken vital monsoon rainfall – study". *E&E News.* 6th November. Available at: https://www.eenews.net/stories/1059990029 (Accessed 21 December 2018).

George, S. 2010. *Whose Crisis, Whose Future?* Cambridge: Polity.

GDR 2018. http://gdrights.org/ Accessed June 14th.

Georgescu-Roegen Nicolas 1971. *The Entropy Law and the Economic Process,* Cambridge MA: Harvard University Press.

Global Commission on the Economy and Climate. 2014. *Better Growth, Better Climate: The New Climate Economy Report.* Washington: World Resources Institute. http://static.newclimateeconomy.report/wp-content/uploads/2014/08/BetterGrowth-BetterClimate_NCE_Synthesis-Report_web.pdf

Global Commission on the Economy and Climate. 2018. *Unlocking the Inclusive Growth Story of the 21st Century: Accelerating Climate Action in Urgent Times.* Washington: World Resources Institute. https://newclimateeconomy.report/2018/

Global Green Growth Institute (GGGI). 2014. GGGI Rwanda: Fact Sheet *Climate Resilient Green Cities.* http://www.greengrowthknowledge.org/sites/default/files/GGGI%20 Rwanda%20Fact%20Sheet%20-%20Climate%20Resilient%20Green%20Gities .pdf. (Accessed 19 December 2018.)

Goodchild, Philip. 2007. *Theology of Money.* London: SCM Press.

Goodier 2010, *A Germ of an Idea,* in the University of Chicago Magazine Jul-Aug 2010 available online at http://magazine.uchicago.edu/1008/features/a-germ-of-an-idea.shtml.

Goodwin, J, J. Jasper, and F. Poletta eds. 2001. *Passionate Politics.* Chicago: Chicago University Press.

Gordon, E.V. (ed.) 1937. *The Battle of Maldon* London: Methuen.

Gould, D. 2009. *Moving Politics: Affect, Emotions and Shifting Political Horizons in the Fight Against AIDS*. Chicago: Chicago University Press.

Graeber, David. 2016. "Despair Fatigue". *The Baffler*, March 2016. https://thebaffler.com/salvos/despair-fatigue-david-graeber

Green European Journal (2015) "Peace, Love and Intervention – Green Views on Foreign Policy" Volume 10 March https://www.greeneuropeanjournal.eu/peace-love-and-intervention-green-views-on-foreign-policy/

Green New Deal Group. 2013. A National Plan for the UK: From Austerity to the Age of the Green New Deal. New Weather Institute. http://www.greennewdealgroup.org/wp-content/uploads/2013/09/Green-New-Deal-5th-Anniversary.pdf.

Greene, G., and Silverthorn, B. 2004. The End of Suburbia: Oil Depletion and the End of the American Dream. [Film]. Toronto, Canada: The Electric Wallpaper Company. https://www.imdb.com/title/tt0446320.

Greiner, Jill et al. (2013). 'Seagrass restoration enhances 'blue carbon' sequestration in coastal waters'. PLOS ONE. 8(8). Available at: https://journals.plos.org/plosone/article?id=10.1371/journal.pone.0072469 (Accessed 21 December 2018).

Griffin, S. 1995. *The Eros of Everyday Life: essays of ecology, gender and society*. New York: Anchor Book/Doubleday.

Guardian 2015 23/03/15 at https://www.theguardian.com/sustainable-business/2015/mar/23/growth-inequality-society-commons-commodification-happiness) accessed 23/11/16.

Guardian 2017, "Climate Change to cause humid heatwaves that will kill even healthy people" *TheGuardian*02/08/17 at https://www.theguardian.com/environment/2017/aug/02/climate-change-to-cause-humid-heatwaves-that-will-kill-even-healthy-people.

Guérot, U. 2017. *Why Europe Should be a Republic.*. https://citiesintransition.eu/interview/ulrike-guerot-why-europe-should-be-a-republic.

Hagemann, M., Hendel-Blackford, S., Höhne, N., Harvey, B., Naess, L. O., and Urban, F. 2011. *Guiding climate compatible development User-orientated analysis of planning tools and methodologies: Analytical Report*. Cologne, Brighton, Utrecht, London: Ecofys and IDS. https://opendocs.ids.ac.uk/opendocs/bitstream/handle/ 1234 56 789/ 4324/ Ecofys%20%20IDS%202011_TM%20for%20CCD_main%20report %20format.pdf?sequence=1&isAllowed=y.

Hajer, M. et al 2015. "Beyond Cockpit-ism: Four Insights to Enhance the Transformative Potential of the Sustainable Development Goals", *Sustainability* 7: 1651–1660.

Hamilton, C. 2010. *Requiem for a species: why we resist the truth about climate change*. London: Earthscan.

Handrich, L., Kemfert, C., Mattes, A., Pavel, F. and Traber, T. (2015) *Turning point: Decoupling Greenhouse Gas Emissions from Economic Growth*. Heinrich Böll Stiftung, Berlin.

Hanson, S., Nicholls, R., Ranger, N., Hallegatte, S., Corfee-Morlot, J., Herweijer, C., and Chateau, J. 2011. "A global ranking of port cities with high exposure to climate extremes." *Climatic Change*. Vol.104:1 (2011): 89–111. https://link.springer.com/content/pdf/10.1007/s10584-010-9977-4.pdf.

Haque, A.N., Grafakos, S and Marijk Huijsman. 2012. "Participatory integrated assessment of flood protection measures for climate adaptation in Dhaka." *Environment and Urbanization* 24:197 (2012) doi: 10.1177/0956247811433538. http://eau.sagepub.com/content/24/1/197.

Hardin, G. 1968. "The Tragedy of the Commons" *Science*, 162 1243–1248.

Harvey, David. 2013. *A Companion to Marx's Capital Volume 2*. London: Verso.

Hayek, Friedrich von. 1976. *Law, Legislation and Liberty, Vol. 2: The Mirage of Social Justice*. London: Routledge.

Head, L. 2016. *Hope and grief in the anthropocene. Reconceptualising human nature relations*. London: Routledge.

Heatley 2015, Green House Gas *Paris: Optimism, Pessimism and Realism*, available online at http://www.greenhousethinktank.org/uploads/4/8/3/2/48324387/paris_-_final2.pdf, and reprinted as Brian Heatley (2017) Paris: optimism, pessimism and realism, Global Discourse, 7:1, 10–22, DOI: 10.1080/23269995.2017.1300402

Heatley B. and R. Read (2017) *Facing up to Climate Reality* http://www.greenhousethinktank.org/uploads/4/8/3/2/48324387/intro_final.pdf (accessed 9 August 2017)

Heilbroner, Robert L. 1985. *The Nature and Logic of Capitalism*. New York: W.W. Norton.

Hickman, L. 2010 "James Lovelock: Humans are too stupid to prevent climate change", *The Guardian* 29th March Available at: https://www.theguardian.com/science/2010/mar/29/james-lovelock-climate-change

HM Government. 2017. *Industrial Strategy: Building a Britain fit for the future*. London: HM Government. https://assets.publishing.service.gov.uk/government/uploads/system/uploads/attachment_data/file/664563/industrial-strategy-white-paper-web-ready-version.pdf.

HM Government. 2018. *The Oxford-Milton Keynes-Cambridge Growth Corridor: Building a Shared Economic Vision*. Accessed December 19 2018. http://cambridgeshirepeterborough-ca.gov.uk/assets/Uploads/Oxford-Cambridge-Growth-Corridor-A-Shared-Economic-Vision-Consultancy-Brief-3.pdf.

HM Treasury. 2014. *National Infrastructure Plan 2014*. London: HM Treasury. https://www.gov.uk/government/publications/national-infrastructure-plan-2014.

Hobsbawm E.,1994, *Age of Extremes*, London, Penguin.

Hoggett, P. 2009. *Politics, Identity and Emotion*. Boulder, Col: Paradigm Publishers.

Hoggett, P. and R. Randall. 2018. "Engaging with climate change: Comparing the cultures of science and activism." *Environmental Values*, 27: 223–243.

Hölderlin, Friedrich 1951. *Sämtliche Werke (Band 2)*, Stuttgart: Kohlhammer.

Holpuch, Amanda. 2018 "Hurricane Maria: Puerto Rico raises official death toll from 64 to 2,975" *The Guardian,* 28th August. https://www.theguardian.com/world/2018/aug/28/hurricane-maria-new-death-toll-estimate-is-close-to-3000. Accessed 27 September 2018.

Homer-Dixon, T. 1999. *Environment, scarcity, and violence.* Princeton: Princeton University Press.

Homer-Dixon, T. 2006. *The upside of down: Catastrophe, creativity, and the renewal of civilisation.* Washington, DC: Shearwater.

Honig, B. 1996. "Difference, dilemmas and the politics of home." In *Democracy and Difference: Contesting the Boundaries of the Political,* edited by S.Benhabib. Princeton, NJ: Princeton University Press.

Hopkins, R. 2008. The Transition Handbook. Totnes: Green books.

Huizinga, J., 1924, *The Waning of the Middle Ages,* English translation. London: Edward Arnold.

Humphrey, S. (ed.) 2009. *Climate Change and Human Rights* Cambridge: CUP.

Hunt, A., P, Watkiss. 2011. "Climate change impacts and adaptation in cities: a review of the literature." *Climatic Change.* Vol 104; No.1; 2011, 13–49. Bath: University of Bath. https://researchportal.bath.ac.uk/en/publications/climate-change-impacts-and-adaptation-in-cities-a-review-of-the-l

Illich, I. 1973. *Tools for Conviviality.* London: Marion Boyars.

IMF. 2015. World Economic Outlook: Adjusting to Lower Commodity Prices. Washington: IMF. www.imf.org/external/pubs/ft/weo/2015/02

IPCC 2001, Intergovernmental Panel on Climate Change, *Third Assessment Report, Part 1 the Scientific Basis,* Cambridge University Press.

IPCC 2007. Fourth Assessment Report: Climate Change AR4. Geneva: IPCC. www.ipcc.ch/publications_and_data/publications_and_data_reports.htm#1.

IPCC 2014a. *Climate Change 2014 Impacts, Adaptation, and Vulnerability Part A: Global and Sectoral Aspects. Working Group II Contribution to the Fifth Assessment Report of the Intergovernmental Panel on Climate Change.* New York, NY: Cambridge University Press.

IPCC 2014b. *Climate Change 2014 Mitigation of Climate Change. Working Group III Contribution to the Fifth Assessment Report of the Intergovernmental Panel on Climate Change.* New York, NY: Cambridge University Press.

IPCC 2014c. IPCC Fifth Assessment Report, Summary for Policymakers. Available at https://ipcc.ch/pdf/assessment-report/ar5/wg3/ipcc_wg3_ar5_summary-for-policymakers.pdf (accessed 23 November 2016).

IPCC 2018. *Special Report on the impacts of global warming of 1.5°C* available at https://www.ipcc.ch/site/assets/uploads/sites/2/2018/07/SR15_SPM_High_Res.pdf

Isildar, G. 2012. Introduction to environmental ethics. In *Environmental ethics: an introduction and learning guide,* edited by K.Vromans, R. Paslack, G. Isildar, R. de Vrind and J. Simon. London: Routledge.

Jackson, T. 2009 / 2011. (2nd ed. 2017) *Prosperity without Growth: Economics for a Finite Planet* London: Earthscan.

Jackson, T. 2018. *The Post Growth Challenge: Secular stagnation, inequality and the limits of growth.* CUSP Working Paper No 12.. https://www.cusp.ac.uk/themes/aetw/wp12/ (Accessed 19 December 2018).

Jackson, Tim, and Peter A. Victor. 2015. "Does Credit Create a 'growth Imperative'? A Quasi-Stationary Economy with Interest-Bearing Debt". *Ecological Economics* 120 (December): 32–48. doi:10.1016/j.ecolecon.2015.09.009.

Jamieson, D. 2014. *Reason in a Dark Time* Oxford: Oxford University Press.

Jasper, J. 1998. "The emotions of protest: Affective and reactive emotions in and around social movements." *Sociological Forum,* 13 (3): 397–424.

Jubilee Debt Campaign. 2018. "Climate scheme hands $105 million profit to global insurance industry". https://jubileedebt.org.uk/press-release/climate-scheme-hands-105-million-profit-to-global-insurance-industry. (Accessed 19 December 2018).

Kanner, A. D., and Gomes, M. E., 1995. "The all-consuming self". In: *Ecopsychology: Restoring the earth, healing the mind,* edited by T. Roszak, M.E. Gomes and A. D. Kanner, 77–91. San Francisco: Sierra Club Books.

Keen, Steve, 2001, *Debunking Economics,* London, Zed Books.

Keen, Steve. 2009. "The Dynamics of the Monetary Circuit". In *The Political Economy of Monetary Circuits: Tradition and Change in Post-Keynesian Economics,* edited by Sergio Rossi and Jean-francois Ponsot, 2009 edition, 161–187. Basingstoke: Palgrave Macmillan.

Kemp, Roger. 2016. *Living Without Electricity: One City's Experience of coping with loss of power.* London: Royal Academy of Engineering.

Kennedy, C. and J. Corfee-Morlot. 2012. Mobilising Investment in Low Carbon, Climate Resilient Infrastructure, Environment Working Paper 46. Paris: OECD.

King, Andrew and B. Henley 2016., *We Have Almost Certainly Blown the 1.5-Degree Global Warming Target,* at http://www.desmog.uk/2016/08/18/we-have-almost-certainly-blown-1-5-degree-global-warming-target (accessed 23 November 2016).

Kishore, N. *et al.* 2018. "Mortality in Puerto Rico after Hurricane Maria", *The New England Journal of Medicine,* May 29. https://www.nejm.org/doi/full/10.1056/NEJMsa1803972,

Klein, N. 2007. *The Shock Doctrine* London: Penguin.

Klein, N. 2014. *This Changes Everything: Capitalism vs. the Climate* Harmondsworth: Penguin.

Klinenberg, Eric. 2015. *Heat Wave: A Social Autopsy of Disaster in Chicago.* 2nd ed. University of Chicago Press.

Klinenberg, Eric. 2016. 'Want to Survive Climate Change? You'll Need a Good Community' *Wired,* November issue. https://www.wired.com/2016/10/klinenberg-transforming-communities-to-survive-climate-change/

Klugman, J. 2009. Human development report 2009. *Overcoming barriers: Human mobility and development.* New York: UNDP. http://hdr.undp.org/sites/default/files/reports/269/hdr_2009_en_complete.pdf

Knight, K. and Schor, J. 2014. Economic Growth and Climate Change: A Cross-National Analysis of Territorial and Consumption-Based Carbon Emissions in High-Income Countries. *Sustainability,* 6, 3722–3731.

Kniveton, D., M, Martin. and P, Rowhani. 2013 *Sensitivity testing current migration patterns to climate change and variability in Bangladesh.* Working paper 5. Accessed December 19 2018. http://migratingoutofpoverty.dfid.gov.uk/files/file.php?name=wp5-ccrm-b-sensitivity.pdf&site=354.

Koistinen, P. 1980. *The Military-Industrial Complex: a historical perspective,* New York: Praeger.

Kumar, K. 1991. *Utopianism* Buckingham: Open University Press.

Lambert, Greg. 2016. "Two months that have been wetter than wet." *Lancaster Guardian,* 7th January.

Lasch. C. 1979. *Culture of Narcissism: American life in an age of diminishing expectations.* New York: W.W. Norton & Company.

Lavoie, Marc. 2015. *Post-Keynesian Economics: New Foundations.* Paperback ed. reprinted with amendments. Cheltenham: Elgar.

Lawn, Philip. 2011. 'Is Steady-State Capitalism Viable?: A Review of the Issues and an Answer in the Affirmative'. *Annals of the New York Academy of Sciences* 1219 (1): 1–25. doi:10.1111/j.1749-6632.2011.05966.x.

Lawson, Nigel, 2008. *An Appeal to Reason: A Cool Look at Global Warming,* Duckworth Over-look, London.

Lear, J. 2006. *Radical Hope: Ethics in the Face of Cultural Devastation* Cambridge MA: Harvard University Press.

Le Bon, G. 1896. *The Crowd: A Study of the Popular Mind.* London: T.F.Unwin.

Le Grand, Julian, and Saul Estrin, eds. 1989. *Market Socialism.* Oxford: Clarendon Press.

Le Roy, Alice. 2017. *The 2003 Paris heat wave and its aftermath in France.* Presentation to 'Dealing with Extreme Weather' conference, Lancaster, 28 October. Available at https://www.greenhousethinktank.org/dealing-with-extreme-weather.html.

Leeson, P. and Everett, E. 2017. *Managing our Uplands.* Presentations to 'Dealing with Extreme Weather' conference, Lancaster, 28 October. Available at https://www.greenhousethinktank.org/dealing-with-extreme-weather.html.

Leopold, A. 1949. *A Sand County Almanac and Sketches Here and There.* Oxford: Oxford University Press.

LeQuesne, C. 1996. *Reforming World Trade: The Social and Environmental Priorities.* Oxford: Oxfam Publishing.

Lertzman, R. 2015. *Environmental Melancholia: Psychoanalytic Dimensions of Engagement*. London: Routledge.

Li, B. 2013. "Governing urban climate change adaptation in China." *Environment and urbanisation*. 2013 25: 413. UK: Sage. https://doi.org/10.1177/0956247813490907.

Li, Minqi. 2007. *Capitalism with Zero Profit Rate? Limits to Growth and the Law of the Tendency for the Fate of Profit to Fall*. Working Paper, University of Utah, Department of Economics.

Lipsey, R. and A. Chrystal 1999. *Principles of Economics*, Ninth Edition, Oxford, OUP.

Lohmann, L. 2006. *Carbon Trading: a Critical Conversation on Climate Change, Privatisation and Power*. Dorset: The Corner House.

Lomborg, Bjorn, 2001 *The Skeptical Environmentalist: Measuring the Real State of the World*, CUP,.

Lonsdale, K., Pringle, P. and Turner, B. 2015. *Transformative adaptation: what it is, why it matters and what is needed*. Oxford: Oxford University, UK Climate Impacts Programme. https://ukcip.ouce.ox.ac.uk/wp-content/PDFs/UKCIP-transformational-adaptation-final.pdf.

Luthar, S., D. Cicchetti, and B. Becker. 2000. "The construct of resilience: A critical evaluation and guidelines for future work." *Child Development* 71 (3): 543–562.

Luxemburg, Rosa. 1964. *The Accumulation of Capital*. London: Routledge.

Lynas, Mark 2011. *The God Species: How the Planet Can Survive the Age of Humans*. London: Fourth Estate.

Machmuller M. B. et al, 2015. "Emerging land use practices rapidly increase soil organic matter". *Nature communications*. 6. Article no. 6995.

MacNeil, William H 1976. *Plagues and People*, Garden City, N.Y.: Anchor Press.

Macy, J. 1993. *World as Lover, World as Self*. Berkeley: Parallax Press.

Macy, J. and C. Johnstone 2012. *Active Hope: How to Face the Mess We're in without Going Crazy* Novato, Cal.: New World Library.

Magdoff, Fred, and John Bellamy Foster. 2011. *What Every Environmentalist Needs to Know about Capitalism: A Citizen's Guide to Capitalism and the Environment*. New York: Monthly Review Press.

Mahmud, W., S, Ahmed. and S, Mahajan. 2008. "Economic Reforms, Growth, and Governance: The Political Economy Aspects of Bangladesh's Development Surprise" in *Leadership and Growth: Report of the World Bank Commission on Growth and Development*, 227. https://openknowledge.worldbank.org/bitstream/handle/10986/24 04/527080PUB0lead101Official0Use0Only1.pdf?sequence=1.

Malthus, Thomas, 1798, *An Essay on the Principle of Population*, 1970 Edition by Penguin Books, London.

Mann, Geoff, and Joel Wainwright. 2018. *Climate Leviathan: A Political Theory of Our Planetary Future*. London: Verso.

Mann, M. E. *et al.* 2017. "Influence of Anthropogenic Climate Change on Planetary Wave Resonance and Extreme Weather Events" *Sci. Rep.* 7, 45242; doi: 10.1038/srep45242 (2017).

Marshall, George 2015. *Don't Even Think About It: Why Our Brains Are Wired to Ignore Climate Change*. New York & London: Bloomsbury.

Marshall, Michael, 2018. "Trouble from the Deep" *New Scientist,* 4 August 2018.

Marshall, Michael. 2010. "Frozen jet stream links Pakistan floods, Russian fires" *New Scientist,* 10 August 2010.

Mason, M. 2005 *The New Accountability: Environmental Responsibility Across Borders.* London: Earthscan.

Mason, Paul 2015, Mason,, *Postcapitalism, A Guide to Our Future,* London: Allen Lane.

McDonald, M. 2013 "Discourses of Climate Security", *Political Geography,* 33: 43–51.

Midgley, M. 2003. *Myths We Live By.* London: Routledge.

Mir, G.U.R. and Storm, S. 2016. *Carbon Emissions and Economic Growth: Production – based versus Consumption-based Evidence on Decoupling.* Institute for New Economic Thinking Working Paper Series.

Missing Pathways to 1.5°C: The role of the land sector in ambitious climate action (2018). Climate Land Ambition and Rights Alliance. October. Available at: https://static1.squarespace.com/static/5b22a4b170e802e32273e68c/t/5bc3cbf28165f51c6af2c7de/1539558397146/MissingPathwaysCLARAexecsumm_2018.pdf (Accessed 21 December 2018)

Mohaupt, S. 2008. "Review article: Resilience and social exclusion." *Social Policy and Society,* 8 (1): 6 – 71.

Monbiot, G. 2017. *Out of the Wreckage: A New Politics for an Age of Crisis* London: Verso.

Mundy, J. 1991, *Europe in the High Middle Ages 1150–1309,* Harlow UK, Longman.

Nalam, Aditya et al. 2018. "Effects of Artic geoengineering on precipitation in the tropical monsoon regions". *Climate Dynamics.* 50(9–10): 3375–3395.

NEF 2009. *Other worlds are possible: Human progress in an age of climate change* The Sixth report of the Working Group on Climate Change and Development London: NEF.

New Scientist staff and Press Association (2016). "World is set to warm 3.4°C by 2100 even with Paris climate deal". *New Scientist.* 3rd November. Available at: https://www.newscientist.com/article/2111263-world-is-set-to-warm-3-4c-by-2100-even-with-paris-climate-deal/ (Accessed 21 December 2018).

New, M., D. Liverman., H. Schroder and K. Anderson. 2011. "Four degrees and beyond: the potential for a global temperature increase of four degrees and its implications." *Phil. Trans. R. Soc. A* 2011 369, 6–19. https://royalsocietypublishing.org/doi/pdf/10.1098/rsta.2010.0303

Newell, P. and M. Paterson 1998. "Climate for business: global warming, the State and capital". *Review of International Political Economy* 5(4), pp. 679–704.

Newell, P. 2000. *Climate for Change: Non-State Actors and the Global Politics of the Greenhouse* Cambridge University Press.

Newell, P. 2001. "Managing Multinationals: The Governance of Investment for the Environment", *Journal of International Development*, 13, 907–919.

Newell, P. 2004. "Climate change and development: A tale of two crises" *IDS Bulletin* (special issue on Climate Change and Development) Vol.35 No.3.

Newell, P. and D. Mulvaney 2012. "The Political Economy of the Just Transition". *The Geographical Journal* 179 (2): 132–40.

Newell, P. and M. Paterson. 2010. *Climate Capitalism: Global Warming and the Transformation of the Global Economy.* Cambridge: Cambridge University Press.

Newell, P. and O. Taylor 2018. "Contested landscapes: The Global Political Economy of Climate Smart Agriculture", *Journal of Peasant Studies* 45(1): 80–88.

Newell, P. and R. Lane 2018. "A climate for change? The impacts of climate change on energy politics" *Cambridge Review of International Affairs*, DOI: 10.1080/0955 7571.2018.1508203

Nixon, R. 2011. *Slow Violence and the Environmentalism of the Poor* Harvard: HUP.

Norberg-Hodge, Helena and Rupert Read 2016. *Post-Growth Localisation.* Local Futures & Green House. Available at: http://www.greenhousethinktank.org/uploads/4/8/3/2/48324387/post-growth-localisation_pamphlet.pdf [accessed 20 December 2018].

Norberg-Hodge, Helena 2008. *Localisation and climate change.* Paper presented at the Climate Change and the Role of Civil Society conference, Seoul, South Korea, 2008.

Norman, Joseph et al. 2015. "Climate Models and Precautionary Measures". *The Black Swan Report.* Available at: https://www.blackswanreport.com/blog/2015/05/our-statement-on-climate-models/ (accessed 21st December 2018).

Nunan, F. (ed.) 2017. *Making Climate Compatible Development Happen* Abingdon: Routledge.

O'Connor, M. (ed.) 1994. *Is Capitalism Sustainable? Political Economy and the Politics of Ecology* New York, Guilford Press.

OECD 2011. *Towards Green Growth*, Organisation for Economic Co-operation and Development, Paris.

OECD 2015. Green Investment Banks: Policy Perspectives. Paris: OECD. http://www.oecd.org/environment/cc/Green-Investment-Banks-POLICY-PERSPECTIVES-web.pdf (Accessed 19 December 2018).

Oil Change International 2016. *Stop funding fossil fuels: World Bank group funds fossil fuel exploration despite calls for climate action.* Available at http://priceofoil.org/content/uploads/2016/04/World-Bank-Brief-April-2016-FINAL2.pdf (accessed 8 August 2017).

Paris Agreement 2015. *United Nations*. Available at: https://unfccc.int/files/essential_background/convention/application/pdf/english_paris_agreement.pdf (Accessed 21 December 2018).

Paul, Helena 2011. "Biochar Knowledge Gaps". *Econexus*. Available at: https://www.econexus.info/publication/biochar-knowledge-gaps (Accessed 21 December 2018).

Peters, K.; Vivekananda, J. 2014. *Topic Guide: Conflict, climate and environment*. London: DFID. doi: 10.12774/eod_tg.november2014.peterskandvivekanandaj. https://www.gov.uk/dfid-research-outputs/topic-guide-conflict-climate-and-environment.

Polanyi, Karl. 2001. *The Great Transformation*. Boston: Beacon Press.

Pollin, Robert. 2018. "De-Growth vs a Green New Deal". *New Left Review*, II,, no. 112: 5–25.

Ponting Clive, 1991, *A Green History of the World*, London, Penguin Books.

Postrel, Virginia. 1990. 'The Green Road to Serfdom'. *Reason.com*, April. https://reason.com/archives/1990/04/01/the-green-road-to-serfdom.

Price Waterhouse Coopers. 2015. *Road Map for Low Carbon and Climate Resilient Kolkata*. Kolkata, India: British Deputy High Commission. Accessed December 19 2018 . https://cdkn.org/wp-content/uploads/2017/07/Up-Low-Carbon-Roadmap-Kolkata-1.pdf.

Proctor, Jonathan et al 2018. "Estimating global agricultural effects of geoengineering using volcanic eruptions". *Nature*. 560: 480–483. Available at: https://www.nature.com/articles/s41586-018-0417-3 (Accessed 21 December 2018).

Proyect, Louis. 2016. "Debates within Ecosocialism: John Bellamy Foster, Jason Moore and CNS". *Louis Proyect: The Unrepentant Marxist*. https://louisproyect.org/2016 / 0 6/15/debates-within-ecosocialism-john-bellamy-foster-jason-moore-and-cns/.

Rabinowitz, Abby and Amanda Simson 2017. "The Dirty Secret of The World's Plan to Avert Climate Disaster". *Wired*. 12th October. Available at: https://www.wired.com/sto r y/ the-dirty-secret-of-the-worlds-plan-to-avert-climate-disaster/ (Accessed 21 December 2018).

Radford, Tim 2017. "Geoengineering Could Create More Problems Than It Could Solve". *Eco Watch*. 24th November. Available at: https://www.ecowatch.com/geoengineering-more-harm-than-good-2512157242.html (Accessed 21 December 2018).

Randall, R. 2013. "Great Expectations: The Psychodynamics of Ecological Debt." In *Engaging with Climate Change: Psychoanalytic and Interdisciplinary Perspectives*, edited by S. Weintrobe, 87–102. Hove, UK: Routledge.

Rapley John, 2017, *Twilight of the Money Gods: Economics as a Religion and How it all Went Wrong*, Simon & Schuster 2017. Summary article in the Guardian at https://www.theguardian.com/news/2017/jul/11/how-economics-became-a-religion.

Raworth, Kate 2017. *Doughnut Economics: Seven Ways to Think Like a 21st-Century Economist*. Manhattan: Random House.

Read, Rupert 2012. *Guardians of the Future: A Constitutional Case for representing and protecting Future People*, Green House report.

Read, Rupert 2015. "Climate science is to geo-engineering as genetics is to GM food". *Respublica*. 30th November. Available at: https://www.respublica.org.uk/disraeli-room -post/2015/11/30/climate-science-geo-engineering-genetics-gm-food/ (Accessed 21st December 2018).

Read, Rupert 2016. "Precaution vs Promethean: The philosophical dividing line that will define 21st century politics". *Rupertread.net*. 4th July. Available at: https://rupertread.net/ precautionary-principle/precaution-vs-promethean-philosophical-divid- ing-line-will-define-21st (Accessed 21st December 2018).

Read, Rupert 2017. "On preparing for the great gift of community that climate disas- ters can give us" in *Global Discourse*. 7(1): 149–167.

Read, Rupert and Deepak Rughani 2017. "Apollo-Earth: A Wake up Call in Our Race against Time". *The Ecologist*. 9th March. Available at: https://theecologist.org/2017/mar/ 09/apollo-earth-wake-call-our-race-against-time (Accessed 21 December 2018).

Read, Rupert 2018. *A Film-Philosophy of Ecology and Enlightenment* Abingdon: Roou- tledge.

Reich, Robert B. 2009. *Supercapitalism: The Battle for Democracy in an Age of Big Busi- ness*. London: Icon Books.

Rio Declaration on Environment and Development 1992. The United Nations Con- ference on Environment and Development. Available at: http://www.unesco.org/ education/pdf/RIO_E.PDF (Accessed 10 January 2019).

Robinson, Joan. 1971. *Economic Heresies: Some Old-Fashioned Questions in Economic Theory*. London: Macmillan.

Robinson, N.A. 2001. "David Ross Brower and Nature's Laws." in *Pace Environmen- tal. L. Rev.* 221(2001). https://digitalcommons.pace.edu/cgi/viewcontent.cgi?arti- cle=1373&context=lawfaculty

Rockstrom, J. et al. (2009) "A safe operating space for humanity" *Nature*. 461.7263 (Sept. 24, 2009): pp 472–5.

Rodionova, Z. 2016. "Housing crisis: architect Bill Dunster designs £50,000 pod homes on stilts for would-be buyers." in *Independent* April 20th, 2016. https://www.independent. co.uk/news/business/news/housing-crisis-architect-bill-dunster-designs-50000-pod -homes-on-stilts-for-would-be-buyers-a6992546.html.

Rouncivell, Gayle. 2016. "City Shops and firms fight back after storms." *Lancaster Guardian*, December 1.

Rust, M. 2008. "Climate on the couch." *Psychotherapy and Politics International*, 6(3): 157–170.

Sassen, S. 2012. *Cities in a World Economy*. London: Sage.

Sassen, S. 2014. *Expulsions: Brutality and Complexity in the Global Economy*. Cam- bridge, MA: Harvard University Press.

Schandl, H., Fischer-Kowalski, M., West, J., Giljum, S., Dittrich, M., Eisenmenger, N., Geschke, A., et al. 2017. "Global Material Flows and Resource Productivity: Forty Years of Evidence". *Journal of Industrial Ecology* 22(1). DOI: 10.1111/jiec.12626

Schmelzer, M. 2016. *The Hegemony of Growth: The OECD and the Making of the Economic Growth Paradigm.* Cambridge University Press.

Schwartz, S. H., J. Cieciuch, M. Vecchione, E. Davidov, R. Fischer, C. Beierlein, A. Ramos, et al. 2012. "Refining the Theory of Basic Individual Values." *Journal of Personality and Social Psychology* 103 (4): 663–688. doi:10.1037/a0029393.

Scott Cato, M. 2012. *The bioregional Economy: land, liberty and the pursuit of happiness.* London: Routledge.

Scott Cato, M. (n/d). *Response from Green House, the environmental think tank.* Environmental Audit Committee Inquiry into the Green Economy.

Selby, J. 2014. "Positivist climate conflict research: A critique" *Geopolitics* 19(4): 829–856.

Semal, Luc, and Mathilde Szuba. 2017. *The "Jacksonisation" of the French Degrowth Discourse* presented at The Politics of Sustainable Prosperity conference, Keele University, July 11. https://www.cusp.ac.uk/themes/p/workshop_11july/.

Seto K.C., S. Dhakal, A. Bigio, H. Blanco, G.C. Delgado, D. Dewar, L. Huang, et al. 2014. "Human Settlements, Infrastructure and Spatial Planning." in *Climate Change 2014: Mitigation of Climate Change.* Contribution of Working Group III to the Fifth Assessment Report of the Intergovernmental Panel on Climate Change [Edenhofer, O., R. Pichs-Madruga, Y. Sokona, E. Farahani, S. Kadner, K. Seyboth, A. Adler, et al (eds.)]. Cambridge: Cambridge University Press.

Shadsuddoha M., S, Khan., S, Raihan., and T, Hossain. 2012. *Displacement and Migration from Climate Hot-Spots: Causes and Consequences.* Dhaka: Center for Participatory Research and Development and Action Aid. http://www.actionaidusa.org/sites/files/actionaid/displacement_and_migration.pdf

Shaikh, Anwar, 2016, *Capitalism: Competition, Conflict and Crisis,* New York, OUP.

Shamsuddoha, M., and R.K., Chowdhury. 2009. *Climate Change Induced Forced Migrants: in need of dignified recognition under a new Protocol.* Dhaka: Equity and Justice Working Group Bangladesh. http://www.mediaterre.org/docactu,Q0RJLUwtMy9kb2NzL2NsaW1hdGUtbWlncmFudC1wcmludGVkLXBvc2l0aW9uLWRlY2l0.1.pdf.

Shapiro, S.L. and Schwartz G.E.R., 1999. "Intentional systemic mindfulness: an integrative model for self – regulation and health." *Advances in Mind-Body Medicine.* 15: 128–134.

Sheldon, K. M., and T. Kasser. 2008. "Psychological threat and extrinsic goal striving." *Motivation and Emotion,* 32: 37–45.

Simms, A. 2013. *Cancel the Apocalypse: The New Path to Prosperity.* London: Little, Brown.

Simms, A. 2008. *Nine Meals from Anarchy: The vulnerability of our oil-dependent society and recommendations about how we might rebuild resilience.* London: New Economics Foundation. https://neweconomics.org/2008/11/nine-meals-anarchy/.

Simms, A. and P, Newell. 2018. *How did we do that? The possibility of rapid transition.* Brighton: Steps Centre. https://steps-centre.org/wp-content/uploads/2017/04/How_Did_We_Do_That.pdf.

Simms, A. and P. Newell 2018. "We need a fossil fuel non-proliferation treaty – and we need it now", *The Guardian* October 23rd Available at: https://www.theguardian.com/commentisfree/2018/oct/23/fossil-fuel-non-proliferation-treaty-climate-breakdown

Smith, J. 2010. *Biofuels and the Globalization of Risk: The Biggest Change in North-South Relationships Since Colonialism?* London: Zed Books.

Smith, Richard. 2015. *Green Capitalism. The God That Failed.* World Economics Association. https://www.worldeconomicsassociation.org/library/green-capitalism-the-god-that-failed/.

Solnit, R. 2004 / 2016. *Hope in the Dark: Untold Histories, Wild Possibilities* Chigago, Il.: Haymarket.

Solnit, R. 2010. *A Paradise Built in Hell: The Extraordinary Communities That Arise in Disaster.* New York: Penguin.

Solo mo n, S., J.L. Greenberg, and T.A. Pyszczynski. 2004. "Lethal Consumption: death-denying materialism". In: *Psychology and Consumer Culture*, edited by T. Kasser and A.D. Kanner, 127–146. Washington DC: American Psychological Association.

Spangenberg, Joachim H. 2010. "The Growth Discourse, Growth Policy and Sustainable Development: Two Thought Experiments". *Journal of Cleaner Production* 18 (6): 561–566. doi:10.1016/j.jclepro.2009.07.007.

Spratt, David 2010. "What would 3 degrees mean?" *Climate Code Red.* 1st September. Available at: http://www.climatecodered.org/2010/09/what-would-3-degrees-mean.html (Accessed 21 December 2018).

Steffen et al. 2015. "Planetary Boundaries: Guiding human development on a changing planet". *Science* Vol. 347 no. 6223 and http://www.stockholmresilience.org/research/research-news/2015-01-15-planetary-boundaries---an-update.html.

Steinbrecher, Ricarda and Helena Paul 2017. "New Genetic Engineering Techniques: Precaution, Risk, and the Need to Develop Prior Societal Technology Assessment". *Environment: Science and Policy for Sustainable Development.* 59(5): 38–47.

Stern, N 2007. *The Stern Review: The Economics of Climate Change.* Cambridge: Cambridge University Press.

Stern, N. 2013. "The structure of economic modelling of the potential impacts of climate change: grafting gross underestimation of risk onto already narrow science models." *Journal of Economic Literature.* Vol.51:3 (2013): 838–859.

Stevenson, H. and J. Dryzek 2014. *Democratizing Global Climate Governance* Cambridge: CUP.

Stirling, A. 2011. "Pluralising progress: From integrative transitions to transformative diversity", *Environmental Innovation and Societal Transitions*, 1(1): 82–88.

Stirling, A. 2014 *Emancipating Transformations: From Controlling 'the Transition' to Culturing Plural Radical Progress*, STEPS Working Paper 64, STEPS Centre, Brighton.

Stirling, A. 2015. *Time to Rei(g)n back the Anthropocene* STEPS blog, October 16th https:// steps-centre.org/blog/time-to-reign-back-the-anthropocene/ (Accessed 5 March 2018.)

Strefler, Jessica et al. 2018. "Potential and costs of carbon dioxide removal by enhanced weathering of rocks". *Environmental Research Letters*. 13(3).

Sutherland, Scott. 2017. "Droughts, floods from weird jet stream linked to Arctic heat." *The Weather Network*. https://www.theweathernetwork.com/news/articles/droughts-floods-from-weird-jet-stream-linked-to-arctic-heat/80841.

Swim, J.K., P.C. Stern, T.J. Doherty, S. Clayton, J.P. Reser, E.U. Weber, R. Gifford. and G.S. Howard. 2011. "Psychology's contributions to understanding and addressing global climate change." *American Psychologist*, 66: 241–250.

Swyngedouw, E. 2010. "Apocalypse forever? Post-political populism and the spectre of climate change", *Theory, Culture and Society*, 27(2–3): 213–232.

Szönyi, Micael, Peter May and Rob Lamb 2016. *Flooding after Storm Desmond*. JBA Trust and Zurich Insurance Group http://www.jbatrust.org/wp-content/uploads/2016/08/flooding-after-storm-desmond-PUBLISHED-24-August-2016.pdf

Taleb, Nassim et al. 2014. *The precautionary principle* NYU Extreme Risk Initiative Working Paper. Available at: http://www.fooledbyrandomness.com/pp2.pdf (Accessed 21 December 2018).

Theisen, O.M. 2017. "Climate Change and Violence: Insights from Political Science", *Current Climate Change Reports* 3(4): 210–211.

Tolstoy, L. 1849/2009. *War and Peace*, trans. Rosemary Edmonds London: Penguin.

Totton, N. 2011. *Wild Therapy: undomesticating inner and outer worlds*. Ross-on-Wye: PCCS Books.

Trainer, T. 1996. *Towards a Sustainable Economy: the Need for Fundamental Change* Oxford, Jon Carpenter.

TUC. 2011. *Briefing: The Green Investment Bank and the Green Economy*. https://www.tuc.org.uk/research-analysis/reports/tuc-briefing-green-investment-bank-and-green-economy. (Accessed 19 December 2018.)

Turner, Adair. 2018. "After the Crisis, the Banks Are Safer but Debt Is a Danger". *Financial Times*. https://www.ft.com/content/9f481d3c-b4de-11e8-a1d8-15c2dd1280ff.

UK Government. 2011. *Climate Resilient Infrastructure: Preparing for a Changing Climate*. https://assets.publishing.service.gov.uk/government/uploads/system/uploads/attachment_data/file/69269/climate-resilient-infrastructure-full.pdf.

UNCC 2015, UNCC, FCCC/CP/2015/7, *Synthesis Report on the aggregate effect of the intended nationally determined contributions*, 30 October 2015.

UN DESA. 2017. *World Population Prospects: the 2017 revision.*. https://www.un.org/development/desa/en/news/population/world-population-prospects-2017.html. (Accessed 19 December 2018.)

UNEP 2015. *Options for decoupling economic growth from water use and water pollution*. Report of the International Resource Panel Working Group on Sustainable Water Management.

UNEP 2016. United Nations Environment Programme, The Emissions Gap Report 2016: A UNEP Synthesis Report http://capacity4dev.ec.europa.eu/unep/document/emissions-gap-report-2016-unep-synthesis-report#sthash.ks4WPfy3.dpuf (accessed 22 November 2016).

UNSC 2007 *Security Council holds first ever debate on impact of climate change on peace, security hearing over 50 speakers*, UN Security Council 5663rd meeting, 17 April 2007; http://www.un.org/News/Press/docs/2007/sc9000.doc.htm

van Oldenborgh, G. J., Otto, F. E. L., Haustein, K., and Cullen, H. 2015. "Climate change increases the probability of heavy rains like those of storm Desmond in the UK – an event attribution study in near-real time". *Hydrol. Earth Syst. Sci. Discuss., 2015*, 13197–13216. doi:10.5194/hessd-12-13197-2015.http://www.hydrol-earth-syst-sci-discuss.net/12/13197/2015/.

Verger, Jacques 1986. "Different Values and Authorities" in Fossier, Robert, (ed.), *The Cambridge Illustrated History of The Middle Ages, Vol 3*, Cambridge, CUP.

Victor, Peter, 2008 *Managing Without Growth: Slower by Design, Not Disaster* Northampton MA: Elgar.

von Weizsäcker, E.U., de Larderel, J.A., Hargroves, K., Hudson, C., Smith, M.H. and Enríquez, M.A. 2014. *Decoupling 2: technologies, opportunities and policy options.* United Nations Environment Programme, Nairobi.

Wallace-Wells, David 2018a. "The Paris Climate Accords Are Looking More and More Like Fantasy".*Intelligencer*.25th March.Available at:http://nymag.com/daily/intelligencer/2018/03/the-paris-climate-accords-are-starting-to-look-like-fantasy.html (Accessed 21 December 2018).

Wallace-Wells, David 2018b. "Solar Geoengineering May Be Our Last Resort for Climate Change. What If It Doesn't Work?". *Intelligencer*. 8th August. Available at: http://nymag.com/intelligencer/2018/08/solar-geoengineering-climate-change.html?gtm=top&gtm=bottom (Accessed 21st December 2018).

Wanner, T. 2015. "The new 'Passive Revolution' of the green economy and growth discourse: Maintaining the 'sustainable development' of neoliberal capitalism". *New Political Economy, 20*(1), 21–41.

Ward J., Sutton P., Werner A., Costanza R., Mohr S., Simmons C. 2016. "Is Decoupling GDP Growth from Environmental Impact Possible?" *PLoS ONE* 11(10)

Ward, H. 2002. *Corporate Accountability in Search of a Treaty? Some Insights from Foreign Direct Liability* Briefing Paper No.4 Chatham House. London: RIIA.

Watts, Jonathan. 2018. "Summer weather is getting 'stuck' due to Arctic warming" *The Guardian,* 20th August. https://www.theguardian.com/environment/2018/aug/20/summer-weather-is-getting-stuck-due-to-arctic-warming?utm_source=esp&utm_medium=Email&utm_campaign=Guardian+Today+-+Collection&utm_term=283858&subid=7555430&CMP=GT_collection.

Weinstein, N. and R.M. Ryan.2011. "A Self-determination Theory Approach to Understanding Stress Incursion and Responses." *Stress and Health,* 27: 4–17

Weinstein, N., K.W. Brown, and R.M. Ryan. 2009. "A multi-method examination of the effects of mindfulness on stress attribution, coping, and emotional well-being." *Journal of Research in Personality,* 43: 374–385

Weintrobe, S. 2004. "Links between grievance, complaint and different forms of entitlement." *Int. J. Psycho-Anal,* 85 (1): 83–96.

Weintrobe, S., ed. 2013. *Engaging with Climate Change: Psychoanalytic and Interdisciplinary Perspectives.* Hove, UK: Routledge.

Welzer, H. 2012. *Climate Wars: Why people will be killed in the 21st century* Cambridge: Polity.

White, L. 1967. "The historical roots of our ecological crisis." *Science,* 155 (3767): 1203–1207.

Wikipedia 2016, *https://commons.wikimedia.org/wiki/File:2000_Year_Temperature_Comparison.png* (accessed 24 October 2016).

Wikipedia 2016, Uneconomic Growth, https://en.wikipedia.org/wiki/Uneconomic_growth (accessed 23 November 2016).

Wikipedia 2018a. https://en.wikipedia.org/wiki/Medieval_Warm_Period (accessed 20 March 2018).

Wikipedia 2018b. https://en.wikipedia.org/wiki/Little_Ice_Age (accessed 20 March 2018).

Wikipedia 2018c. https://en.wikipedia.org/wiki/Medieval_demography (accessed 20 March 2018).

Wilde, O. 1891 / 1973. "The Soul of Man under Socialism" in *De Profundis and Other Writings,* ed. H. Pearson. Harmondsworth: Penguin.

Williams, B. 1981. *Moral Luck: Philosophical Papers 1973–1980.* Cambridge University Press.

Williams, R. 1961. *The Long Revolution.* London: Chatto & Windus.

Wilson, Edward 2016. *Half-Earth: Our Planet's Fight for Life.* New York & London: Liveright Publishing Corporation.

World Bank 2010 *Development and Climate Change.* World Development Report. Washington: The World Bank.

World Bank (2012) *Inclusive Green Growth: The Pathway to Sustainable Development,* World Bank, Washington DC.

World Bank. 2018. *Groundswell – Preparing for Internal Climate Migration*. Washington: World Bank. https://openknowledge.worldbank.org/handle/10986/29461.

World Bank. 2018. *Nigeria Economic Update: Connecting to Compete*. Accessed 19 December 2018. https://www.worldbank.org/en/news/press-release/2018/05/02/nigeria-economic-update-connecting-to-compete.

Wright, C. and D. Nyberg. 2015. *Climate change, capitalism, and corporations: processes of creative self-destruction*. Cambridge: Cambridge University Press.

Wright, R. 2004. *A Short History of Progress*. Edinburgh: Canongate Books.

WWF. 2012. *Living Planet Report 2012*. Woking: WWF. http://wwf.panda.org/knowledge_hub/all_publications/living_planet_report_timeline/lpr_2012/.

Yeampierre, Elizabeth and Klein, Naomi 2017. "Imagine a Puerto Rico Recovery designed by Puerto Ricans" *The Intercept*. https://theintercept.com/2017/10/20/puerto-rico-hurricane-debt-relief/ (accessed 20 December 2018).

SPONSORS

Green House would like to thank the following who have generously made substantial donations to assist in the publication of this book:

- Ian Christie
- David Hirst
- The Polden-Puckham Charitable Foundation
- John Ranken
- Margaret Robinson
- Paul Vare